Praise for Promise Ahead

"[Promise Ahead] alert[s] us to important problems and offers sugges-tion[s] that are genuinely constructive."

—*Christian Science Monitor*

"The very act of envisioning a better future can in itself initiate change. . . . *Promise Ahead* inspire[s] those necessary dreams."

—New Age

"Essential reading for people flooded with information yet filled with confusion. . . . Duane Elgin has the courage and the credentials to provide us with a visionary and plausible road map into the deep future."

—Vicki Robin, coauthor of *Your Money or Your Life*

"A compelling blueprint for the future that is both hopeful and doable. This book deserves the attention of our entire society."

—Larry Dossey, M.D., author of
Reinventing Medicine and *Healing Words*

"Well-reasoned and plausible . . . . Duane Elgin is calling us on a hero's journey where we can all be heroes."

—Robert Johansen, president, Institute for the Future

"Essential reading for those dedicated to creating a world that works for all."

—David C. Korten, author of
The Post-Corporate World and *When Corporations Rule the Worl*

"*Promise Ahead* is a book for the twenty-first century with a clear message: a sustainable planet is not an unreachable goal."

—Elizabeth Dowdeswell, former director,
United Nations Environment Program

"Well-researched and reasoned and offers us a rare treasure-grounds for hope."

—Sam Keen, author of
Fire in the Belly and *To a Dancing God*

"*Promise Ahead* offers us a new framework for our conscious evolution."

—Barbara Hubbard, author of
Conscious Evolution and *The Evolutionary Journey*

Also by Duane Elgin

Voluntary Simplicity

Awakening Earth

Promise Ahead

A Vision of Hope and Action for Humanity's Future

Duane Elgin

Quill
An Imprint of HarperCollins*Publishers*

A hardcover edition of this book was published in 2000 by William Morrow, an imprint of HarperCollins Publishers.

First Quill edition published 2001.

Designed by Jam Design

The Library of Congress has catalogued the hardcover edition as follows:

Elgin, Duane.
Promise ahead : a vision of hope and action for humanity's future / by Duane Elgin.
p. cm.
Includes bibliographical references (p.) and index.
ISBN 0-688-17191-5
1. Philosophical anthropology. 2. Social evolution. I. Title.
BD450.E397 2000
128—dc21 99-087191

ISBN 0-06-093499-9 (pbk.)

03 04 05 BP 10 9 8 7 6 5

Dedicated to my mother and father,

Mary and Clifford Elgin,

with love and appreciation for

the kindness and creativity

they demonstrated in their lives

Contents

Introduction

by *Vicki Robin*

SOME books require no introduction. Within pages a good novel has you in the grip of its plot and characters. How-to books assume that you suffer from the complaint of the day and quickly lure you in with the promise of material salvation through their particular plan. Even many of the recent spate of spiritual books fit into this category. Contemporary nonfiction books, from the political to the prurient, require only that you have followed the news to draw you in.

All of these books, though, tell pretty much the same familiar story. We live short and unpredictable lives, struggling in a world that is often senseless and cruel. Islands of goodness, from romance to family to business success, can take the edge off this reality. So can the millions of consumer products touted from every television screen and billboard. The steady stream of scandals that greets us at the checkout counter of the supermarket week in and week out keeps us equally mesmerized. People who find themselves anchored in some haven of security may tithe a token of their time or money to ease the burden of "those less fortunate." But the story remains—poverty of every sort will always be with us. Politics—from family to national—is the art of making do.

Isn't this true? Isn't the very repetitiveness of this story, however dreary, somehow comforting? How many of us want to be disturbed by the kind of idealism that rises up in human groups from time to time, hinting that life itself might promise much, much more than thin thighs and fat wallets?

This book, *Promise Ahead,* invites you into that arena of grounded idealism, into the world of dreaming a new dream not just for your personal life, but for a multifaceted, rich and finely detailed unfolding story of our whole species. While this might not seem to matter to your day-to-day life, the promise of such a future can rearrange your personal world in quite remarkable ways.

Not too long ago, when John Lennon and the Beatles invited us to "imagine" a sweet, gracious, and peaceful world, we had the audacity to allow ourselves to dream. Now we have to wade through layers of distraction and demands to gain the ease that imagining requires. So let me invite you to briefly remove your twenty-first-century sophistication and indulge in some very pleasant fantasies.

Imagine that your boss likes your work. Imagine that your responsibilities are such that you can truly clear your desk and close up shop at the end of the day. Imagine that your house is truly a home, a haven of peace, that your commute is short and that you've just paid off your last debt (yes, even your house and car). Imagine that your kids like each other/school/what you fix for dinner and get decent grades. Imagine you have enough time to follow the thread of their curiosity about why things are the way they are. Imagine that every day something happens that makes you smile. It could happen, admit it.

Imagine trusting the media and the government again.

The news informs. People care. Politicians are public servants and make a median wage. Imagine . . .

We've arrived, now, at a very special place. The world of possibility. The unknown. Out of the unknown will come everything of real value in your life, because the future is, in reality, unknown. Your children and grandchildren, your next jobs and eventual retirement, your vacations and new friends are all waiting for you—in the unknown. Imagination is like a steering wheel for this world of infinite possibility. If you start to shed the quiet despair about the possibility of your life and our collective life ever making sense again, you might just find yourself with childlike eagerness, peering expectantly into the unknown.

Entering the space of imagination will help you enter the magnificent world that Duane Elgin shares in *Promise Ahead*—not because it's a book of fantasy, but rather because it's a book based in years of research and contemplation about achingly beautiful possibilities embedded in hard reality. He surveys much of what's known about the universe and our place in it, then invites us to peer into the unknown with him and imagine an evolutionary journey that's better than any Hollywood pyrotechnics could evoke.

In *Promise Ahead* Elgin invites us to think in several new ways: First, we are invited to approach living as a member of the human species. Sure, we have our personal story, abundant with friends, family, activities, and significance. But we are also part of a larger identity, the "body" of humanity. As a species, as humanity, we have a history much longer than a few measly decades. You and I are actually thirty-five thousand years old. We've invented tools, language, cities, and civilization. We've puzzled over the mysteries of life and created stories and religions to link us to the unseen world. We've mastered fire, directing it to warm our homes, power our cars, and send us into space. At the same time, our bombs and guns, designed to subdue our enemies, are returning to haunt us as children shoot children and terrorists reduce buildings to rubble. All of us, with all the achievements and contradictions of our species, are part of this human journey. While thinking of oneself as part of humanity is no shocking revelation, thinking that humanity itself can, as a single creature, think—that's the stretch you're being invited to take. What if "humanity" is more than just a designation that distinguishes our species from other creatures? What if "humanity," like an individual, is on a journey—and is at a crossroads? Elgin asks us to contemplate with awe the beauty and terror of our collective pilgrimage through time and space, take stock and choose, *both consciously and collectively,* our future.

Second, we are invited to see this mighty task in a unique way. Our collective history of social, cultural, and political change has been traditionally presented as an ongoing struggle and clash of ideas and people, all vying for power over the resources of the present and the directions of the future. Elgin, though, sees where we are and where we are headed through a different lens. For him, our historical power dynamics can be seen as kid's stuff—the "terrible two's," the skinned knees of grade school, the teenage recklessness that is the stuff of parental nightmares. As a species, we've just been kids in the cosmos—making mistakes, making mischief, making friends, and making more of us at an astounding rate. Now we are at that turning point called "growing up." Will we, like Peter Pan, refuse to mature? Or will we, *as a species,* have the will, good sense and courage to move on to adulthood?

"Growing up" for many teens has about as much appeal as a bath does for a dog. Don't you know a lot of teenagers running around in adult bodies, defying limits everywhere—overspending, speeding, playing around on their mates, and using various drugs to mute their consciences? But consider traditional cultures (and our own a few short generations ago). Achieving adulthood was more like winning an Olympic gold medal. We endured many trials to prove to our elders we were worthy to be counted as one of them. This is the opportunity humanity now faces, according to Elgin. Growing up, in the best sense of the phrase.

While optional, this choice to mature is by no means window dressing. It is very consequential. Adults, by their very nature, *want* to understand and nurture the world beyond the boundaries of their own self-centered playpen. With all the social and environmental challenges ahead, we need a wisdom crew on Spaceship Earth, not a bunch of unattended children amusing themselves with expensive and dangerous toys.

Third, with the nature of change itself changing, Elgin says our hope lies in the simple power of conscious communication, not in traditional forms of analysis and organizing. As humanity bonds with itself and together faces the future, we'll need to do what all marriage counselors recommend: talk with one another. Talking, though, doesn't mean just chatter. It means purposefully bringing up those tough subjects we'd all rather ignore, listening to opinions that don't match our own, thinking clearly, speaking accurately, and—most amazing—acting on new information or insights. Just because we started to talk at age two doesn't mean we know how to communicate. This learned art, hard enough by itself, is getting harder by the day as we pour the oatmeal of junk-information all over the wiring of the global brain—the media. Elgin singles out the imperative to communicate intelligently via this collective voice as one of humanity's essential next steps. Enriching the menu of options on traditional media is certainly essential to upping our collective IQ, but the intoxicating wild card is the internet. How we use this precious gift of connectivity can steer our species out of the shallows of mediocrity and into our true brilliance. Our news must broaden again out of the constraints of infotainment, our discussions must foster respect and insight, and our democracy, drowning now in cynicism and consumerism, must actually start to work. We need good information, great conversations, and a sense that our voices can be heard. Our collective brain must hum with the aliveness of millions of bits of accurate data shuttling around, resonate with the pure drumbeat of feelings, crackle with enlightening insights, and be bathed in the water of compassion. To put it simply, we need a good head on our shoulders. So it will be from the stuff of dialogue, not ideology, that the future will be made.

Finally, Elgin invites our imagination (but not our incredulity) to expand into the vast reaches of space and time. He doesn't ask us to take any leaps of faith that have no basis in science. Rather, he lays before us what science has unearthed about our more-than-earthly reality. He explores recent findings in physics that point to the possibility that our universe is a single, living system and may not be all that exists "out there." Taken together, these insights reverberate with meaning. Our lives—including our most mundane decisions—are part of a coherent, purposeful unfolding. Yet nothing is assured. *We* must wake up to our personal and social wholeness and act like . . . well . . . grown-ups!

We seem to be on the brink of as big a shift in our collective understanding of the cosmos as people faced back in the "flat earth" days. Those flat-earthers, though, had several centuries to make the shift, whereas all of us alive are headed into this new reality at breakneck speed. We are walking—no, racing—into the unknown together. With courage, imagination, and knowledge, we can embrace this mysterious wind that is blowing in from the future. We can enjoy the journey and thrill at the *Promise Ahead*. Whaddya say? Are you on board?

Vicki Robin is the coauthor with Joe Dominguez of *Your Money or Your Life*.

Chapter One

Is Humanity Growing Up?

Life is occupied both in perpetuating itself
and in surpassing itself;
if all it does is maintain itself,
then living is only not dying.
—*Simone de Beauvoir*

Humanity's Age

HOW grown up do you think humanity is? When you look at human behavior around the world and then imagine our species as one individual, how old would that person be? A toddler? A teenager? A young adult? An elder?

As I've traveled in different parts of the world, speaking to diverse audiences, I've begun many of my presentations by asking this question. Initially, I didn't know whether people would be able to relate to or even understand my question, much less agree on an answer. To my surprise, nearly everyone I've asked has understood this question immediately and has had an intuitive sense of the human family's level of maturity. Whether I've asked this question in the United States, England, India, Japan, or Brazil, within seconds people have responded in the same way: *at least two-thirds say that humanity is in its teenage years.*

The speed and consistency with which different groups around the world have come to this intuitive conclusion were so striking that I began to explore adolescent psychology. I quickly discovered that there are many parallels between humanity's current behavior and that of teenagers:

- Teenagers are *rebellious* and want to prove their independence. Humanity has been rebelling against nature for thousands of years, trying to prove that we are independent from it.
- Teenagers are *reckless* and tend to live without regard for the consequences of their behavior. The human family has been acting recklessly in consuming natural resources as if they would last forever; polluting the air, water, and land of the planet; and exterminating a significant part of animal and plant life on the Earth.
- Teenagers are concerned with *appearance* and with fitting in. Similarly, many humans seem focused on expressing their identity and status through material possessions.
- Teenagers are drawn toward instant *gratification.* As a species, we are seeking our own pleasures and largely ignoring the needs of other species and future generations.
- Teenagers tend to gather in groups or *cliques,* and often express "us versus them" and "in versus out" thinking and behavior. We are often clustered into ethnic, racial, religious, and other groupings that separate us from one another, making an "us versus them" mentality widespread in today's world.

Other authors have noted that we are acting like teenagers. Al Gore wrote in his book, *Earth in the Balance,* "The metaphor is irresistible: a civilization that has, like an adolescent, acquired new powers but not the maturity to use them wisely also runs the risk of an unrealistic sense of immortality and a dulled perception of serious danger. . . ."[1] In a similar vein, Allen Hammond, senior scientist at the World Resources Institute, who has been exploring the world of 2050, has written, "Just as parents struggle to teach their children to think ahead, to choose a future and not just drift through life, it is high time that human society as a whole learns to do the same."[2]

If people around the world are accurate in their assessment that the human family has entered its adolescence, that could explain much about humanity's current behavior, and could give us hope for the future. It is promising to consider the possibility that human beings may not be far from a new level of maturity. If we do develop beyond our adolescence, our species could begin to behave as teenagers around the world do when they move into early adulthood: we could begin to settle down, think about building a family, look for meaningful work, and make longer-range plans for the future.

Adolescence is a time when others—such as parents, schools, churches, and so on—are generally in control. As we step into adulthood, we enjoy a new freedom from control, and a new responsibility to take charge of our lives. In a similar way, during our adolescence as citizens of the Earth, most humans have felt controlled by someone else—especially by big institutions of business, government, religion, and the media. As we grow into our early adulthood as a species, we will discover that maturity requires taking more responsibility and recognizing that we are in charge. Instead of waiting for "Mom or Dad to fix things," an adult pays attention to the larger situation and then acts, recognizing that our personal and collective success are deeply intertwined.

Is it plausible that humanity is truly on the verge of moving beyond our adolescence? Not only do I consider it plausible, I would like to offer a rough timetable for the maturing of humanity. I estimate that we awoke in the *infancy* of our potentials roughly thirty-five thousand years ago. Archeologists have found that, at that time, there was a virtual explosion of sophisticated stone tools, elaborate burials, personal ornaments, and cave paintings. Then, with the end of the ice ages roughly ten thousand years ago, we began to settle down in small farming villages. I believe this period marks the transition to humanity's *childhood*. The food surplus that peasants produced made possible the eventual rise of small cities. I estimate we humans then moved into our *late childhood* with the rise of city-state civilizations roughly five thousand years ago in Iraq, Egypt, India, China, and the Americas. At that time, all the basic arts of civilization were developed, such as writing, mathematics, astronomy, civil codes, and central government. Still, the vast majority of people lived as impoverished and illiterate peasants who had no expectation of material progress. With the scientific-industrial revolution roughly three hundred years ago, humanity began to move into our *adolescence*. Beginning in Europe and the United States, industrialization has spread around the world, particularly in the last half-century. Now, with the industrial revolution devastating the whole planet and challenging humanity to a new level of stewardship, it seems plausible that we are on the verge of moving into the communications era and our *early adulthood*.

This timetable gives only a rough estimate of the average level of maturity of our species, but it does make an important point: that human beings are growing up, becoming more seasoned and wiser through hard-earned experience. Despite humanity's seeming immaturity in the past, I believe we could be close—within a few decades—of taking a major step forward in our evolution as a species.

Humanity's Heroic Journey of Awakening

IF we look beneath the complexity of human history and culture, there seems to be a story that humanity shares regarding the purpose of life. Joseph Campbell, a world-renowned scholar who spent a lifetime exploring the stories that have brought meaning to people throughout history, described the common story at the heart of all the world's cultures as the "hero's journey." Although the details vary depending on where and when it has been told, it is essentially the story of an individual who grows up by going through a series of tests that teach him or her about the nature of life. The person then brings this precious knowledge back to his or her personal life and life of the community.

If we assume that the overall human family is on an heroic journey of development, then the pivotal question becomes: "Where are we on the hero's journey?" To explore that key question, it is important to know that the hero's journey usually consists of three stages: separation, initiation, and return.[3] It begins with the hero (or heroine) leaving home to search for the deeper meaning and purpose of life. This is the stage of separation. There eventually comes a time when the hero undergoes a supreme test, whereby he is initiated into the nature and ways of the universe and no longer feels separate. With initiation, he experiences the deep unity and aliveness at the foundation of the universe and his sense of life purpose in relation to it. He returns from his adventure with that hard-won knowledge and the capacity for personal renewal or even, says Campbell, "the means for the regeneration of society."[4] The core purpose of that sacred knowledge, according to Campbell, is to "waken and maintain in the individual a sense of wonder and participation in the mystery of this finally inscrutable universe."

Just as all major cultures share the story of the hero's journey, all have customs of initiation as well. Initiatory rites of passage around the world have at least two things in common.[5] First, for the individual, the initiation marks a decisive transition from one stage or kind of life to another (such as from adolescence to adulthood). Second, initiation rites are also stressful social situations in which new ways of relating to other people are learned and established. The experience of initiation forges bonds of connection among those who have gone through it, bonds that transcend previous distinctions based on status, age, or kinship. Long after the rites are concluded, these links and emotional bonds persist and provide much of the social glue that holds the community together.[6]

Let's look at humanity's journey in terms of this simple model of separation, initiation, and return.

- **Separation**—By my reckoning, a complex phase of progressively divorcing ourselves from nature has extended over the last thirty-five thousand years—from the time of our initial awakening as gatherers and hunters up to the present. During these millennia, humanity has increasingly pulled away from nature in order to develop our unique capacities and talents as a species. The last half-century seems to mark our final severance from nature as we cause, for example, the mass extinction of other species and the disruption of the global climate.

- **Initiation**—To undergo initiation is to make a major transition to a new and larger life, and it often involves going through a powerful experience. As we confront challenges such as climate change and species extinction, humanity seems poised to undergo an initiation that will give us the opportunity of becoming an authentic human family—in feeling and experience as well as in name.

- **Return**—To pass successfully through our initiation, we will have to forge new bonds both within our species and with nature as a whole. This phase marks our passage into our early adulthood and the beginning of a long process of reconnecting with nature. A promising future lies ahead as we begin a task for grown-ups—building a sustainable planetary civilization.

Figure 1 illustrates this description of our evolutionary journey as separating from and then reconnecting with nature. It is important to note that there is no negative implication in the downward direction of the first arrow. Its purpose is simply to show that we pulled away from nature. We shall soon enter a period of initiation, in which we see that we have a choice of connecting consciously with nature and the universe. Making the transition from separation to integration, without losing the scientific understanding and technical sophistication we have gained, is perhaps the most important evolutionary turn that humanity will ever have to accomplish.

When we view humanity's evolution this way, our times take on new significance. Humanity is about to move into

FIGURE I

Two Major Phases in the Human Journey

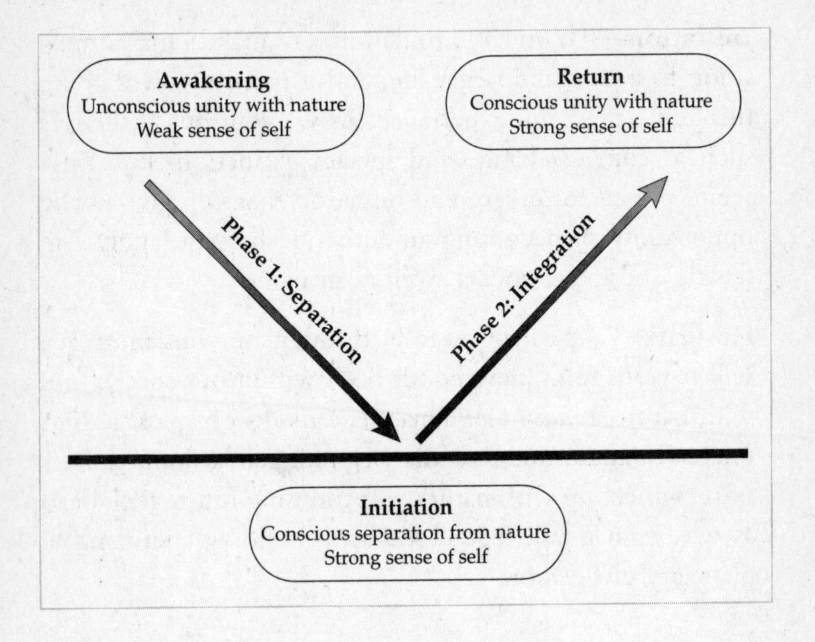

a stage of initiation—a period of stress and testing in which we will be challenged to discover ourselves as a single family with responsibilities to one another, the Earth, and future generations. Although the challenges we face may seem to be evidence of humanity's failures, reaching this stage is actually an expression of our great success over the past thirty-five thousand years. I believe that the apparent crises we face are, in reality, part of our initiation into a new relationship with one another and the Earth. The rapidly approaching initiation represents a time of birth—a stressful but entirely natural process.[7]

When we began our journey of awakening roughly thirty-five thousand years ago, we had only an indistinct sense of ourselves and a strong but largely unconscious feeling of connection with nature.[8] Over the millennia, we have acquired a strong sense of ourselves, but at the cost of separating ourselves from nature. Looking ahead, we have the opportunity to reconnect consciously with nature and the larger human family. As in the hero's journey, our challenge is to return to where we started, but with a new level of insight, compassion, and creativity. T. S. Eliot foretold of this return when he wrote: "And the end of all our exploring/Will be to arrive where we started/And know the place for the first time."[9]

Initiation: Hitting the Evolutionary Wall

HERE is how William D. Ruckelshaus, the former director of the Environmental Protection Agency, describes the evolutionary task that we are facing:

> Can we move nations and people in the direction of sustainability? Such a move would be a modification of society comparable in scale to only two other changes: the Agricultural Revolution and the Industrial Revolution of the past two centuries. Those revolutions were gradual, spontaneous, and largely unconscious. This one will have to be a fully conscious operation . . . If we actually do it, the undertaking will be absolutely unique in humanity's stay on the Earth.[10]

What would motivate us to attempt such an undertaking? I believe it will take both the push of environmental necessity and the pull of evolutionary opportunity for humanity to attempt to overcome thirty-five thousand years of progressive separation from nature and discover how to live in conscious harmony with one another and the Earth. Like adolescents pressing to find the limits of their parents' authority, we are pushing up against the limits of nature, as though seeking to discover just how much abuse our planet will tolerate. But we face much more than physical problems; we face equally great challenges in our own consciousness and character. The historic path of development is being confronted not only with an environmental wall but by an even more formidable evolutionary wall. It would be useful to distinguish here between the two:

- An **environmental wall** refers to the physical limits of the global ecosystem to support our species. We are fast approaching these limits because we are rapidly consuming more resources than the Earth can renew and polluting the environment with more toxins than it can absorb. Given an abundance of resources, nearly every organism exploits its environmental niche to the fullest extent; thus, overshoot and collapse are a common occurrence in natural systems. Human beings learn through experience, and we have no experience exercising restraint as a species and being mindful of the overall biosphere. Since we have never encountered this situation before, it seems only natural that humanity would reach, and then extend beyond, the limits of the Earth's ecosystem.

- An **evolutionary wall** refers not only to the physical limits of the Earth's ability to sustain humanity, but also to our own social and spiritual limits to sustain dysfunctional and destructive behaviors. Modern, industrial civilization is breeding pathological behavior—alienation from others and from nature, extreme competitiveness and greed, cynicism in politics, and despair for the future. How much poverty, alienation, and misery can humanity experience before we eventually damage our collective psyche and soul? An evolutionary wall presents humanity with an identity crisis at least as great as our environmental crisis: Who are we as a species? What is our larger story? Are we going to allow "overshoot and collapse" to happen to us? Do we see ourselves as separate, isolated beings or part of the larger web of life?

In seeing the initiation that awaits us, it is clear that we have come to a great choice-point in our journey. Although human beings have been faced with challenges throughout history, we have never before been confronted with a challenge to our entire planet and species. *Our time is unique in one crucial respect: the circle has closed—there is nowhere to escape.* For the first time in our history, the entire human population is confronted with a common predicament whose solution will require us to work together.

This book looks beyond the possibility of a destructive evolutionary crash to the possibility of an evolutionary bounce. I believe that in the coming decades, there is the distinct possibility that we may surpass ourselves and evolve to a level of maturity that we could not attain without confronting these trials that I am calling "initiation." How might an evolutionary bounce look? I see it as a leap forward in our collective maturity to build a life together that would be harmonious in three ways. It would be:

- **Sustainable**—in harmony with the Earth's biosphere (the *physical* ecosystem)

- **Satisfying**—in harmony with others (the *social-cultural* ecosystem)
- **Soulful**—in harmony with the "life force" (the *spiritual* ecosystem)

There are two compelling reasons for making this evolutionary turn. First, it is eminently desirable and will lead to a higher quality of life for all. Second, it is necessary if we are to avoid creating a planet that is hotter, hungrier, poorer, and more polluted, diseased, and biologically impoverished than it already is.[11]

Humanity's Promising Future

IF we do get through these difficult times and grow into our early adulthood as a species, how long might we then survive? We can gain some perspective by looking at the longevity of early humans and other animal species. The typical life span of a species is estimated to be between one and 10 million years.[12] For example, our early ancestor *Homo erectus* survived more than a million years before becoming extinct. Some species live far longer. Dinosaurs survived roughly 140 million years before a natural catastrophe wiped them out. If humanity is as capable of survival as the dinosaurs were, our species would be able to endure for more than twenty-five thousand times the span of recorded human history. *If we can make it through this evolutionary initiation and begin building a sustainable, satisfying, and soulful planetary civilization, we have the prospect of a long and promising future.*

Just as every child makes missteps on the path to adulthood, humanity has made and will continue to make painful mistakes as we evolve. We learn through our mistakes, however, and we keep moving ahead step by step. We are ever more experienced, ever more seasoned, and ever more mature. Although our future is uncertain, we already have the resources and capacities we need for success. The biologist Lewis Thomas describes the promise of our species beautifully:

> We may all be going through a kind of childhood in the evolution of our kind of animal . . . If we can stay alive, my guess is that we will someday amaze ourselves by what we can become as a species. Looked at as larvae, even as juveniles, for all our folly, we are a splendid, promising form of life and I am on our side.[13]

I too believe that humanity has a promising future. The word *promise* has its origin in the Latin word "to send forth." A promise, then, is a sending forth of a declaration, vow, pledge, or commitment. I believe we are reaching a unique point in our evolution where we can make a promise to future generations. It is a declaration that we will not forget them in the rush and busyness of our day-to-day lives. The promise is our marriage with the larger flow of life—both past and future—and our recognition that we are now a critical link in maintaining the integrity of that flow. It is our sacred covenant with the future whereby we send ahead not only our good intentions, but also our commitment of active engagement to turn the direction of our evolution in favor of a promising future. It is our vow to future generations that we shall hold them in our hearts and minds as we make decisions, recognizing that we all share the same Earth and a common journey through eternity.

Looking Ahead

THE remainder of this book explores humanity's journey toward our young adulthood. We begin in chapter two by taking a hard look at the world and at the key adversity trends that we face. Then in chapters three through six, we explore four equally powerful opportunity trends. Next, we look at the convergence of these trends and consider two basic outcomes—either an evolutionary crash or an evolutionary bounce. In chapter eight, we step back to consider the big picture of the human journey. An evolutionary bounce means the human family will pull together for a common purpose. But what purpose is so compelling that it overcomes historic divisions and differences? Sustainability alone promises little more than "only not dying." Is there a higher purpose that describes our journey? Finally, with perspective for the journey ahead, we return in chapter nine to where we started, to consider how we can collectively awaken and mobilize ourselves to realize the promise ahead.

Chapter Two

Adversity Trends: Hitting an Evolutionary Wall

What is difficult is to imagine how to get out of the situation we're in right now in a time frame that is in line with the rate of deterioration that we're seeing.

—*Paul Hawken*

If we do not change our direction,
we are likely to end up where we are headed.

—*Ancient Chinese Proverb*

Are We on a Collision Course with Nature?

ASSUMING that our species is in its teenage years, I don't think we will easily turn away from our rebellious, reckless, and shortsighted behavior. In fact, we seem determined to run headlong into the consequences of our adolescent actions before deciding whether to make the turn toward a higher maturity. Like many teenagers, we will likely face a time of testing and initiation before moving into our early adulthood.

The question is whether we will pull together as a human family under the extreme pressures of approaching the time of initiation. To answer this question, we need a much clearer sense of the driving trends that will intersect in the next few decades. We begin by considering the problematic trends that promise great misfortune for humanity if we do not face up to them squarely. I call them adversity trends. Two questions seem paramount: How difficult might our situation become? And how soon might we encounter an unyielding evolutionary wall?

After studying driving trends for more than thirty years, I am all too aware that no one can predict the future with certainty. I also know that we can make educated guesses about how the major trends—population, resources, and environment—will unfold in the decades just ahead. Where disagreements emerge is in interpreting the overall meaning of the combined trends. On the one hand, there are some who believe that, with engineering, biotechnology, and human ingenuity, we can solve the problems we face and realize an ever-improving future. On the other hand, there are those who conclude from these same trends that we have already overreached our relationship with life on our planet and, to secure a sustainable future, we will need a profound change in human culture and consciousness as much as a change in technology.

Writing about the "coming age of abundance," Stephen Moore is an economist who epitomizes the perspective of technological optimism: "Every measurable trend of the past century suggests that humanity will soon be entering an age of increasing and unprecedented natural resource abundance."[1] Fred Smith, president of the Competitive Enterprise Institute, writes that while "doomsayers" think there are too many people consuming too much for the planet to sustain, "cornucopians, in contrast, argue that humanity faces no real problems; technological and institutional advances have and will continue to make it possible to address any shortages."[2] The late Julian Simon, a former professor of business, is another optimist: "The standard of living has risen along with the size of the world's population since the beginning of recorded time. There is no convincing economic reason why these trends toward a better life should not continue indefinitely."[3] His rationale for this optimism is that, historically, the opportunity for people to make a profit has spurred human ingenuity and problem-solving—and we end up better off.[4]

The rosy views of the future portrayed by these economists benefit from the fact that, over the last few decades, various predictions of calamity have not materialized. For example, the global "population bomb" was projected to result in massive famines by the turn of the century. As dates for eco-catastrophe and global famine have come and gone without the disastrous events occurring, people's patience for ominous predictions has worn thin.

But can even the most optimistic among us afford to brush off the warnings of the world's leading scientists? In 1992, more than sixteen hundred of the world's senior scientists, including a majority of the living Nobel laureates in the sciences, signed an unprecedented "Warning to Humanity." In this historic statement, they declared that "human beings and the natural world are on a collision course . . . that may so alter the living world that it will be unable to sustain life in the manner that we know." This is their conclusion:

> We, the undersigned senior members of the world's scientific community, hereby warn all humanity of what lies ahead. *A*

great change in our stewardship of the earth and the life on it is required, if vast human misery is to be avoided and our global home on this planet is not to be irretrievably mutilated[5] [emphasis added].

Is this a valid warning? Are we on a collision course with nature and perhaps our own human nature? As a way to explore this vital question, let us look one generation into the future—roughly the next twenty to thirty years—and picture the world that a child born at the turn of the millennium will likely inhabit as a young adult. What kind of legacy are those of us who are alive today leaving for the next generation?

There are dozens of trends that we could consider, such as ozone depletion, rain forest destruction, topsoil erosion, and the overfishing of the world's oceans. To keep the inquiry manageable, however, let's consider only five key driving trends: climate change, population growth, species extinction, resource depletion, and global poverty. These five adversity trends will be sufficient to reveal whether the warning from our leading scientists is valid and, if so, the time frame within which we risk "irretrievable mutilation" of the biosphere.

Global Climate Change

It is no accident that, of the ten warmest years on record, all have occurred in the last fifteen years. In 1995, the Intergovernmental Panel on Climate Change (IPCC)—the international body charged by the United Nations to study global climate change—reached the conclusion that "there is a discernible human influence on global climate."[6] They found that the primary cause for these climate changes is the increase in greenhouse gases that trap heat in the atmosphere. The principal greenhouse gas is carbon dioxide, which comes from burning gasoline, coal, and natural gas. Figure 2 shows how accumulations of carbon dioxide have recently skyrocketed compared to levels during the last hundred thousand years. It also shows how, over the millennia, the rise and fall of global temperatures have corresponded closely with the rise and fall of concentrations of carbon dioxide in the atmosphere. Lastly, this figure suggests that we should expect a major disturbance in the generally favorable weather the world has experienced over the last ten thousand years (since the beginnings of agriculture and a settled way of life).

The several thousand scientists involved in the IPCC study have determined that preindustrial levels of carbon dioxide will at least double by the middle of the next century.[7] This is very bad news, because there is a growing scientific consensus that anything more than a doubling of greenhouse gas concentrations beyond preindustrial levels poses an unacceptable risk.[8] If the levels of atmospheric carbon dioxide should double as the IPCC scientists predict, here are some of the impacts that we could expect:

- Widespread disruption and dislocation of agricultural growing regions
- More rain in some areas, more drought in others
- Stronger storms, more floods, stronger hurricanes
- Stronger effects from El Niño
- Heat waves that kill people, animals, and crops
- Expansion of the Earth's deserts

F I G U R E 2

Global Temperatures and CO_2 Levels

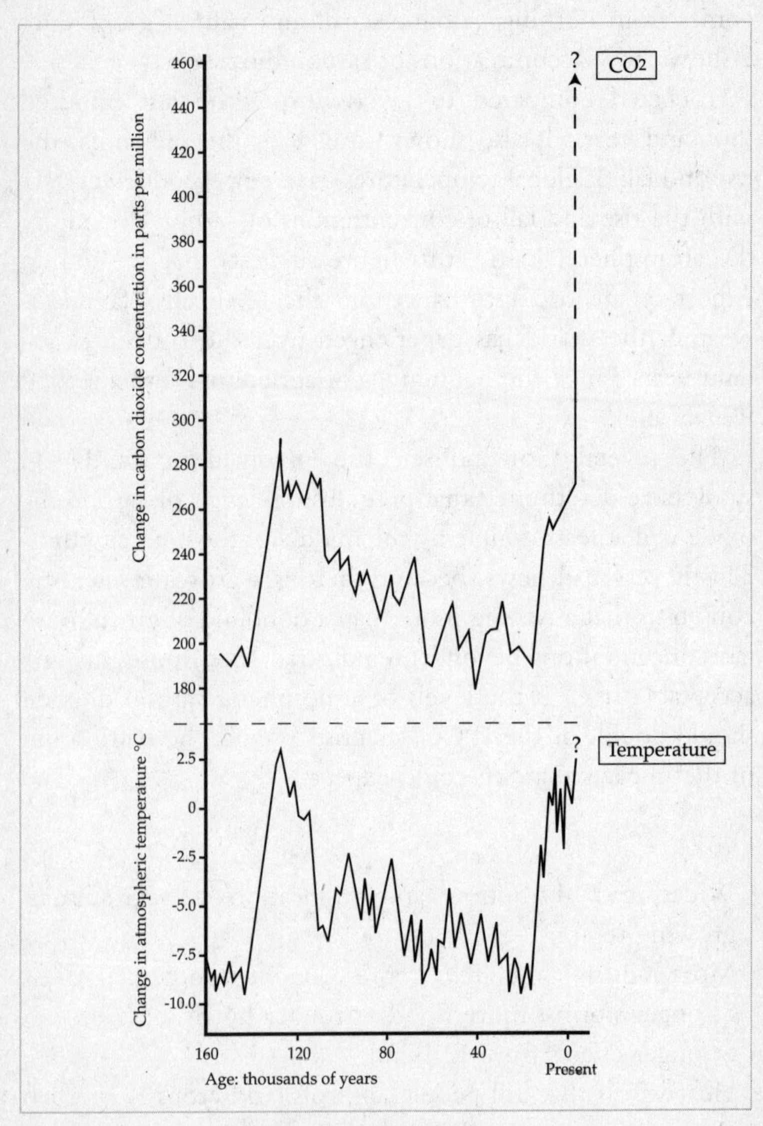

Source: Adapted from Al Gore, *Earth in the Balance*

- Melting of the polar ice caps, with a destructive rise in the sea level
- The spread of infectious diseases that endanger human and animal health
- Stress on the rest of the ecosystem (forests, wetlands, natural habitats)
- Enormous financial burdens placed on individuals, communities, insurance companies, other public and private financial institutions, and nations

The increase in greenhouse gases is already affecting our climate in two distinct ways: it is producing greater variability in weather patterns and increasing the average global temperature. Even now, we are experiencing greater variability in weather patterns. This instability, which is expected to increase considerably, is particularly harmful for agriculture. An early frost in the fall can kill some crops and freeze others in the ground. Heavy rains in the spring can delay planting. A shift in rainfall patterns can change the kind of crops that can be grown in a given area. Because an erratic and unstable climate jeopardizes the productivity of global agriculture, the melting of the global ice caps is not required for the greenhouse effect to have a disastrous impact on the human family.

Here are just a few examples of how global warming might affect agriculture. Canada's climate could improve the area as a wheat-growing region, although its soils are not as productive as the prairie soil in the United States that was built up over millions of years.[9] More frequent droughts in the "breadbasket" of the American Midwest would make it difficult to maintain current levels of productivity in growing wheat and corn. Agricultural productivity will also likely fall in sub-Saharan Africa, parts of Asia, and tropical Latin America—the regions where many of the world's poorest people live. With these kinds of impacts occurring around the world, climate change could cause a dramatic restructuring of the economy at all levels, from local to global.

How rapidly might major changes in the world's climate occur? Scientists are reaching the stunning conclusion that the increase in greenhouse gases is producing a warmer world much faster than expected—so fast that even present generations could feel the dire impacts of global warming. The last great ice age began roughly 120,000 years ago, with a period of abrupt global warming that was followed by rapid cooling. Evidence now indicates that, within a hundred-year span, there was a period of global warming that caused oceans to rise as much as twenty feet, followed by a period of rapid cooling in which the oceans fell by nearly fifty feet. If the human family were to be taken on such a breathtaking roller-coaster ride of massive and sudden climate fluctuations, the consequences would be disastrous.

Instead of there being a gradual warming trend over decades—one to which we could adapt—the world's climate could change abruptly, becoming suddenly much warmer or cooler. For example, dramatic cooling could occur in Europe if the North Atlantic Current—the enormous flow of warm water from the southern hemisphere to the north—were disrupted. Roughly the equivalent of a hundred Amazon Rivers, this current is a conveyor belt of warm water that slowly flows from the equatorial region up to the North Atlantic, giving Europe an unusually warm and favorable climate. Were it not for this flow of warm water, Europe would have the climate of Canada, and its now bountiful agriculture would be reduced to a fraction of current levels. Global warming could bring bring the North Atlantic Current to a halt, with catastrophic consequences for Europe. Professor William Calvin has researched this possibility and has concluded that "the abrupt cooling promoted by man-made warming looks like a particularly efficient means of committing mass suicide."[10]

This first adversity trend—global climate change—is so powerful that it could, by itself, constitute an environmental wall that forces humanity to change directions. Climate change promises to be a persistent force in our collective future as we will run up against its consequences for centuries to come. Although we cannot predict the degree of fluctuation and disruption that will occur in the next several decades, a growing body of research indicates that the changes could be far more rapid and substantial than anyone previously thought.

World Population Growth

CURRENT trends in world population growth give reason for both optimism and concern.[11] There is cause for optimism because we are moving toward global population stability. There is cause for concern because there remains so much momentum in population growth (with so many young people now reaching child-bearing age) that the human family will increase by several billion more people in the decades ahead before we reach stability. Mid-range estimates are that population will grow for another fifty years before it peaks at around 10 billion—four billion more peo-

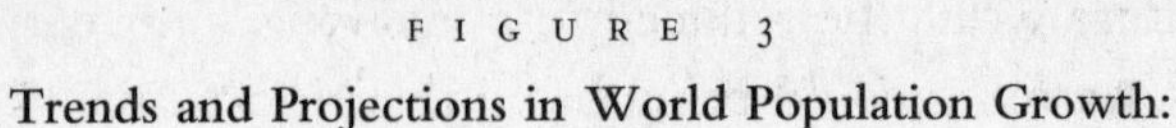

FIGURE 3

Trends and Projections in World Population Growth: 1750–2150 (in billions of people)

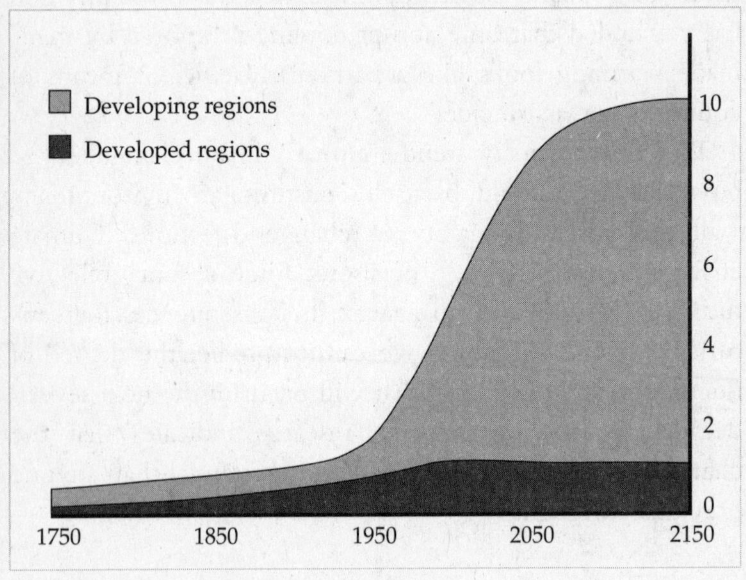

Source: World Resources Institute, 1996–97

ple than the number alive at the year 2000. The predicted range in global population in 2050 is between 8 and 12 billion people, with 10 billion being the mid-range estimate.[12]

Figure 3 shows world population growth in both developed and developing regions from 1750 (the beginning of the Industrial Revolution), with projections to 2150 (this is assuming, of course, that there will be no great die-off in the human population).

Within the time frame covered by this book (roughly until the end of the 2020s), middle-range estimates are that two to three billion people will be added to the Earth's population.[13] This is roughly the equivalent of adding the population of a city the size of New York or Los Angeles to the Earth *each month*. Moreover, 95 percent of this population growth is expected to occur in the poorest countries, which can least afford it, putting enormous pressures on natural resources and already overburdened cities.

It appears that cities, particularly those in developing countries, will bear much of the burden of increasing population. This is because of migration as well as population growth. The movement of people from rural to urban areas is changing the complexion of human culture. A half-century ago, the world was overwhelmingly rural. At the turn of the millennium, the world is half rural and half urban. It is estimated that, by 2050, two-thirds of the world's population will be urban. The shift to a predominantly urban world will produce a radical change in humanity's cultural consciousness.

In the twenty-five years between 1990 and 2015, increasing births and migration will cause the cities in the world's poorest countries to grow by an estimated two billion people.[14] These impoverished cities will then hold three-quarters of the urban population of the world. Poor countries are not ready for this flood of humanity; they are still trying to cope with the one billion people who arrived during the last forty years. The United Nations estimates that by 2015 there will be twenty-seven mega-cities in the world, each with a population of 10 million or more; twenty-three of these cities will be in developing countries. The Worldwatch Institute estimates that, if recent trends continue, 6.5 billion people will live in cities by 2050, more than the world's total population at the turn of the millennium.[15] Yet, even today, urban areas in developing countries contain massive shantytowns that lack adequate housing, paved roads, sewers, clean water, public transportation, schools, health care, fire and police protection, and space to grow food. Because of overcrowding and a lack of sanitation, epidemics of the past—such as cholera, dysentery, and typhoid—are returning. If the U.N. estimates prove to be correct, we will see an enormous burgeoning of these places of misery, pollution, and disease within the next few decades.

It appears that we may be approaching a Malthusian solution to the world population problem. Either by choice (people consciously shifting to smaller families) or by consequence (increasing deaths from starvation, disease, or conflicts over resources), the world population may stabilize by mid-century at around 8 billion people instead of the 10 billion currently predicted. However many people join the human population in the next fifty years, it is clear that humanity is already pushing against the ability of the Earth to carry the burden of our current forms of development. Population growth constitutes a powerful dimension of the evolutionary wall—one that could turn humanity either toward a more sustainable path of development or toward chaos.

Mass Extinction of Species

PERHAPS the most direct measure of the health of the biosphere is the status of its biological diversity. Instead of supporting a flourishing and robust biosphere, humans are busy cutting down forests, overfishing the oceans, paving over the land, and pouring toxins into the water, soil, and air. The net result is decimation of the community of plant and animal life on the Earth. The health of the planet is in jeopardy as industrial activity is causing mass extinction of animal and plant species.

In a 1998 survey of four hundred scientists by the American Museum of Natural History in New York City, nearly 70 percent of the biologists polled said they believed that a *"mass extinction" is under way, and predicted that up to one-fifth of all living species could disappear within 30 years"*[16] [emphasis added]. They attribute nearly all the loss of plant and animal species to human activity. Another estimate—this one appearing in *National Geographic* magazine—concludes that 50 percent of the world's plant and animal species (fish, birds, insects, and mammals) could be on a path to extinction within a hundred years.[17]

Such massive extinctions have occurred only five times before, and each of these was due to natural causes such as the impact of an asteroid and sudden climate change. The sixth great extinction is now under way, and we are the cause. Because of the ruthless way in which humans are destroying other living organisms and their habitats, we have been called an exterminator species.

While it is true that mass extinctions have occurred before and life on the planet has recovered, it is also true that recovery was very slow—generally over millions of years. The mass extinction that is now under way is progressively degrading the resilience and integrity of the biosphere. As plants and animals disappear, their absence affects the entire ecosystem, particularly with regard to natural services such as pollination, seed dispersal, insect control, and nutrient cycling.[18] In addition, with a smaller pool of species, there are fewer candidates to take the place of those species that cannot weather catastrophic droughts, freezes, pest invasions, and diseases.[19] Diminishing biodiversity also affects health care; roughly 25 percent of the drugs prescribed in the United States include chemical compounds from wild organisms.[20] Illustrative of these natural medicines are digitalis, which is derived from foxglove and used to treat heart failure; taxol, a cancer drug that is derived from the bark of the yew tree; and an important blood-clotting agent found in the horseshoe crab. We have no idea how many natural medicines are being eliminated as we clear-cut forests, pollute rivers and oceans, and pave over natural habitats to make room for our enormous cities.

Our extermination of other species has been compared to popping rivets out of the wings of an airplane in flight. How many rivets can the plane lose before it begins to fall apart catastrophically? How many species can our planet lose before we cross a critical threshold where the integrity of the web of life is so compromised that it begins to come apart, like an airplane that loses too many rivets and disintegrates? Loss of biodiversity is another adversity trend that, by itself, could create an unyielding evolutionary wall.

Depletion of Natural Resources

THERE are many natural resources that we are rapidly depleting with no regard for future generations. Two of these vital resources—water and oil—illustrate how we are approaching limits to traditional forms of growth.

Growing water scarcity. Studies by the World Bank have found that worldwide demand for water is soaring and that supply cannot keep pace with demand.[21] In 1995, eighty countries were found to have water shortages that threatened both public health and economic health. More than 40 percent of the world—more than 2 billion people—did not have access to clean water at that time. Northern China, western and southern India, South America, sub-Saharan Africa, and much of Mexico all face water scarcity. In the twenty-first century, the most ferocious competition for resources may not be for oil but for fresh water.

Sandra Postel, who does research on international water and sustainability issues, estimates that, *"by 2025, nearly 40 percent of the world's population will be living in countries whose water supplies are too limited for food self-sufficiency"*[22] (emphasis added). Reasons for the scarcity include overpumping of ground water and the redirection of water from agriculture to cities.

Hundreds of Chinese cities already face acute water shortages—in some, residents are allocated only a meager trickle of water for a few hours each day. Another ominous indicator is the condition of China's famed Yellow River, which supports the breadbasket of China's agriculture. In 1972, it ran dry for the first time in history; since 1985, it has run dry each year until 1997, when it failed to reach the sea for more than half of the year.

Because China depends on irrigation to produce 70 percent of its grain, water shortages translate into food shortages and the need to import grain.[23] Importing a ton of grain is the equivalent of importing a thousand tons of water used to grow that grain elsewhere. China has such an enormous population that its imports of grain could absorb all of the world's available exports, thereby pushing up sharply the world price of grain. For the 1.3 billion people around the world who live at the edge of survival—on the equivalent of $1 or less per day—these price increases could be life threatening. Food shortages could quickly translate into profound social and economic instability around the world.

Former U.S. Senator Paul Simon estimates that by the year 2050, 4 billion people will live in regions of severe water shortages. He writes that "no nation's leaders would hesitate to battle for adequate water supplies," the lack of which "could result in the most devastating natural disaster since history has been recorded . . ."[24]

The end of cheap oil. Much of the growth of the industrial era has been fueled by a one-time gift from nature: the fossil fuels that accumulated over millions of years. Petroleum has fueled not only our transportation but also a revolution in agriculture with petroleum-based pesticides and fertilizers. Although oil prices at the turn of the millennium are quite low, fossil fuel resources are running out more quickly than most people realize.

Colin Campbell and Jean Laherrere have each worked in the oil industry for more than forty years. They predict that conventional oil production will begin to decline by around 2010. "There is only so much crude oil in the world, and the industry has found about 90 percent of it."[25] They conclude that "barring a global recession, it seems most likely that world production of conventional oil will peak during the first decade of the 21st century."[26] Even optimistic projections of remaining reserves suggest that conventional oil will top out by 2020.[27] The respected oil geologist L. F. Ivanhoe agrees with this estimate and recently wrote, "Most of the world's large, economically viable oil fields have already been found, so a permanent oil shock is inevitable early in the next century."[28] The permanent oil shortage will begin, he says, when the world's demand for oil exceeds global production—a condition he expects to emerge around 2010.

There are alternatives to petroleum such as natural gas, solar energy, geothermal energy, wind power, and fuel cells. The widespread use of fuel-cell technology will be particularly important. Nonpolluting and highly efficient, fuel cells create electricity through a chemical reaction that uses gasoline or hydrogen gas and produces no emissions but water. Fuel cells will transform the power systems for automobiles and could be in widespread use by 2015.[29]

If we make the transition to a variety of alternative approaches to energy, there need be no long-term scarcity. It will take decades, however, for us to convert to renewable energy systems, or to develop fuel cell technologies to serve 6 billion people, or to build pipelines for a natural-gas-based economy. We now have several, precious decades in which to make this transition, but little is being done to implement a sustainable energy system globally. Thus, it seems likely that the end of cheap oil will result in serious economic and social dislocations by the 2020s.

An often overlooked consequence of the end of cheap oil will be that the cost of maintaining a high level of agricultural productivity will rise as petroleum-based pesticides, herbicides, and fertilizers become more expensive. At the very time the Earth will contain an added 2 to 3 billion people, the skyrocketing cost of petroleum could undermine the ability of the poorest countries to feed those additional billions. Although high-yield agriculture is possible without heavy reliance on petroleum-based products, it would be a different kind of agricultural system than the one in place today. It would be smaller in scale, decentralized, attentive to local conditions, and operated by well-trained organic farmers. The key question is whether we can make a smooth transition to another agricultural system as the cheap oil runs out in the next decade or two.[30]

Overall, the world water crisis and oil crisis seem to be reaching a critical threshold in roughly the same time frame of the next several decades. While we could respond with foresight—for example, by investing in renewable energy alternatives for oil and desalinization plants for water—we are not yet displaying a high level (or even a modest level) of collective initiative. Therefore, the depletion of critical natural resources such as oil and water constitutes another powerful factor in defining the evolutionary wall.

Poverty and Diminished Opportunity

THE late prime minister of Canada, Lester Pearson, observed: "No planet can survive half slave, half free; half engulfed in misery, half careening along toward the supposed joys of an almost unlimited consumption . . . Neither ecology nor our morality could survive such contrasts."[31] If history is any guide, as the world becomes increasingly divided into the rich and the desperately impoverished, it will produce revolutionary movements in search of fairness and justice. Figure 4 vividly illustrates how far we are from an equitable distribution of material wealth. It shows the percent of global income distributed among five equal segments of the world's population.

If the poverty line is set at $1 a day, at the turn of the millennium it includes 1.3 billion people or roughly 20 percent of humanity. If the poverty line is set at $3 a day, it includes roughly 3.6 billion people or some 60 percent of

FIGURE 4

Global Income Distribution[32]

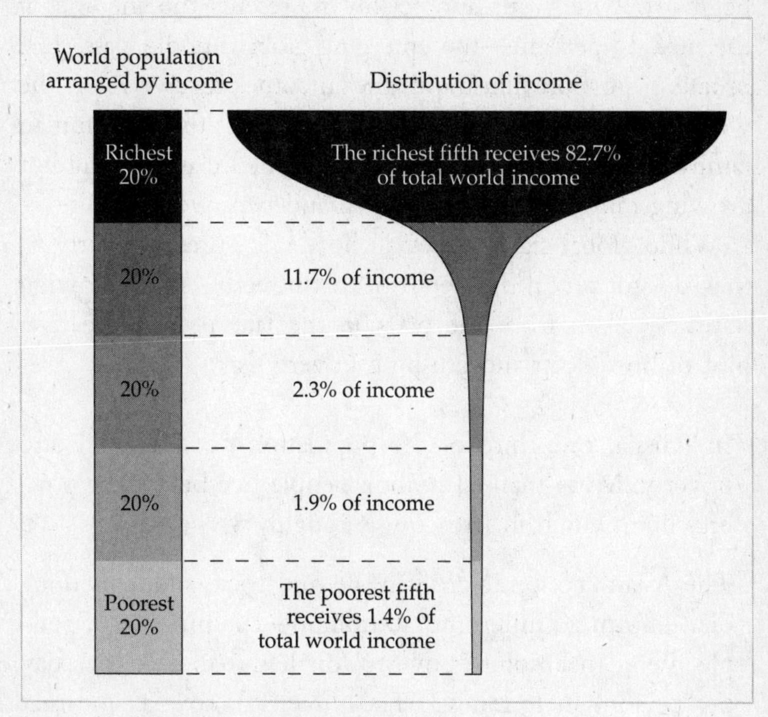

Source: UNDP, *Human Development Report* 1992

humanity. By comparison, the official poverty level in the United States is approximately $11 a day or $4,000 per year, per person—almost four times more than what most people in the world subsist on. What this means is that grinding poverty and the absence of opportunity are the way of life for a majority of human beings today.

Another way to consider global income distribution is by looking at the shape of Figure 4. The long, thin portion of the outline (akin to the stem of a champagne glass) represents the annual income of a majority—approximately 60 percent of the people in the world. The portion where the stem begins to widen out appreciably represents the income of the next 20 percent—the emerging global middle class. The broadest portion illustrates the income received by the world's wealthiest 20 percent. It is apparent that the human family is made up of a huge impoverished class, a small but growing middle class, and a small and very wealthy elite.

While global statistics reveal how widespread poverty is, they do not reveal the depth of that poverty. The following statistics, taken from the press in the late 1990s, give us a hint of how deep and crushing poverty is:

- In Russia, one-third of the population now lives in dire poverty. More than 44 million people live below the poverty line, which in Moscow is roughly $1 per day.[33]
- The Asian economic crisis in the mid 1990s suddenly doubled—from 20 million to 40 million—the number of people living in absolute poverty (on less than a $1 per day per person) in Indonesia. The poverty is so extreme that doctors at two clinics said the number of patients had fallen by half because people could no longer afford to pay the consultation fee, the equivalent of five cents in U.S. currency.[34]
- Although China's economy is growing rapidly, the World Bank estimates that "more than one-quarter of all Chinese—about 350 million—are in substantial deprivation, subsisting on less than $1 a day. Of these, 60 to 100 million are on the edge of starvation with less than 60 cents a day."[35]

- India continues to be the world's poorest nation. More than 500 million Indians earn less than $1 a day (many of them earn less than 10 cents a day) in terms of real purchasing power.[36] Another indication of deprivation: 78 percent of Indian homes do not have access to electricity.[37]

- Worldwide, almost 3 billion people (50 percent of the world's population) use wood as their primary source of energy.

Nothing reveals the vulnerability of the world's poor more than the prospect of widespread food scarcity. As long as global food production grows faster than population, the poor can be relatively patient, hoping that their share of the world's resources will rise eventually.[38] When food production falls behind population growth, however, then how food is allocated becomes an intense and immediate political issue. As Lester Brown has said, while there are substitutes for oil, there are no substitutes for food. He uses the case of China—which contains one-fifth of humanity and has recently become a net food-importing nation—to illustrate the coming food challenge:

> If China's rapid industrialization continues . . . its import demand will soon overwhelm the export capacity of the United States and all other grain-exporting countries. China's rising grain prices are now becoming the world's rising grain prices, China's land scarcity will become everyone's land scarcity, and China's water scarcity will become the world's water scarcity.[39]

Compounding this situation, by the 2020s, billions of people will be living in urban slums without clean water, sanitation, telephones, transportation, health care, or a place to grow food—yet most will have access to television that shows them in vivid detail the high-consumption lifestyles that will never be theirs. This is a recipe for resentment and revolution. For these billions, even a small rise in the price of food can be a serious threat to survival. To have a majority of humanity struggle all day to make a meager living and then view on television a flood of advertisements from the affluent world is to create a schizophrenic planet that is divided against itself. A world in which a majority of people are both wired and poor is a highly unstable world. Fueled by hopelessness and desperation, the potential for terrorism will be enormous.

Although a core concern of this book is the prospect of a whole-systems or planetary-wide crisis, it is important to acknowledge that, for the greater part of the human family, a systems breakdown is a daily reality. A majority of the human family already lives on less than $3 a day in real income, with an impoverished diet, without access to health care, in urban shantytowns and rural villages without electricity, clean water, or fire and police protection. In this sea of poverty and diminished opportunity claiming billions of people, a systems crisis is a current fact.

The human family is in a real quandary. On the one hand, if poverty and famine grow, our collective future will be in doubt as we descend into resource wars and civil breakdown. On the other hand, if consumption grows unrestrained, our collective future will be in doubt as we overtax the Earth's limits.

A World Bank report released in 1997 projects that five

developing countries—Brazil, China, India, Indonesia, and Russia—will become economic superpowers by 2020.[40] Projections are that the economies of China and Indonesia will each grow to more than five times their present size, that of India will grow four times, and that of Brazil will grow three times. Because the Earth is already showing signs of ecological stress from current levels of economic activity, Lester Brown poses the obvious question: "If the global economy is already overrunning its natural capacities, what happens as China, India, and other fast-developing countries strive to emulate the American lifestyle?"[41] For example, if car ownership and oil consumption per person in China were to reach U.S. levels, then China would consume roughly 80 million barrels of oil per day. Yet, in 1996, the entire world produced only 64 million barrels of oil per day. This is a stark example of how the industrial model of development "is not viable for China or for the world as a whole, simply because there are not enough resources."[42]

Our planet does not have the resources to sustain the consumerist culture that is rapidly spreading from developed nations to the rest of the world. The only path through the dilemma of increasing poverty on the one hand and natural limits to unrestrained consumption on the other hand seems to be a middle way of conscious balance and mutually assured development.

The Coming World-System Challenge

WHEN we look a generation ahead to the decade of the 2020s, it is clear that the five adversity trends that we have just considered could dramatically alter the experience of living on the Earth. We have a tendency to compartmentalize these powerful trends and think that we can deal with them one by one when, in reality, they are increasingly interacting and amplifying each other's impact. Here are some of the interconnections that I see. The success of our global industrial economy is destabilizing the climate and supporting an unprecedented increase in world population. A dramatically larger population together with expanding industrialization are consuming an increasing proportion of the Earth's biosphere and destroying ecosystems, causing the most massive extinction of species in the last 65 million years. As industrialization continues, water is becoming scarce and cheap oil is being depleted—and both these resources are vital to agricultural production. Another result of expanding population and industrialization is that several billion human beings now live in impoverished urban slums without access to land and water for growing their own food. At the very time that worldwide agriculture will have increased demands placed on it, it will be taxed by climatic changes, shortages of water and cheap oil (which is needed for fertilizer and pesticides), and possibly even the loss of many of the pollinating insects that are crucial to agricultural production. The result could be waves of famine in regions that do not have enough water and arable land to be self-sufficient. Another result could be resource wars and massive civil unrest.

We do not have to go down this path and hit the wall of absolute limits to growth. We have the time and opportunity to design ways of living that are adapted to the unique ecology, culture, and resources of each bioregion of the planet. But instead of progress, we see continued deterioration and delay. This is described in the 1997 United Nations progress report issued five years after the first Earth Summit in Rio:

> Progress towards a global sustainable future is just too slow. Internationally and nationally, the funds and political will are insufficient . . . to address the most pressing environmental problems even though the technology and knowledge are available to do so. The gap between what has been done thus far and what is realistically needed is widening.[43]

It seems as though our situation is not yet sufficiently critical to mobilize the collective attention of people around the world. It seems likely, however, that we will soon encounter enough pressure from adversity trends to force us to wake up to our time of choice: Will we or will we not collectively take responsibility for the health of the human family and our planet? How we respond to that wake-up call will be a direct measure of our maturity as an evolving species.

WILL WE LEARN THE LESSONS OF HISTORY?

IF the adage is true that those who do not remember the lessons of history are destined to repeat them, then it would be wise for us to recall the civilizations throughout history that have overtaxed their environmental base and collapsed.[44] One was Sumer—one of the earliest known civilizations—that emerged in Mesopotamia, which is in present-day Iraq. Sumerian civilization blossomed in cities in the region around 3,000 B.C. and flourished for more than a thousand years before disintegrating. All that remains today is a desolate and largely treeless landscape. To understand the growth and decline of Sumerian civilization, it is important to examine how the productivity of its agricultural base changed over the centuries.

Despite its hot and dry climate, Sumer achieved a high level of agricultural productivity and a food surplus through extensive irrigation. Irrigation was a stimulus for civilization; it required cooperative efforts and large-scale organization to dig and repair the canals and divide the water. Irrigation also increased the salt content in the soil, a process which eventually undermined the productivity of the land. Crop yields remained high for roughly six hundred years until salinization finally reached critical thresholds and crop yields fell dramatically. From 2000 B.C. onward there are reports that "the earth turned white"—a clear reference to the layer of salt on top of the soil.[45] By 1800 B.C. crop yields had fallen to a third of their initial levels and the agricultural base for Sumerian society effectively collapsed. This contributed to destructive wars between the city-states and exposed Sumer to external attacks that finally overwhelmed it.[46] The Sumerian region then "declined into insignificance as an underpopulated, impoverished backwater of empire."[47]

Other examples of ecological destruction impairing civilizational growth can be found in the Mediterranean region. Vegetation in the Mediterranean is now characterized by low bushes, vines, and olive trees. Before human settlement, however, the natural vegetation of the region was a forest with a diverse mixture of trees. As the land became populated, trees were cleared for agriculture, for building homes and boats, and for cooking and heating. Overgrazing of the cleared land by goats, cattle, and sheep insured that the original forests could not grow back. The signs of widespread environmental destruction were clearly evident in Greece by 650 B.C. Plato left this description of deforestation and soil erosion in his *Critias:* "What now remains compared with what then existed is like the skeleton of a sick man, all the fat and soft earth having wasted away and only the bare framework of the land being left."[48]

The Roman empire was another civilization in which environmental deterioration became an important contributing factor in its decline. As Rome expanded its empire, it placed increasing pressures on the Mediterranean to provide food to support its large urban population as well as its standing armies. These pressures eventually led to widespread deforestation and soil erosion in the lands around the Mediterranean, including northern Africa. The decline in the quality of the soils and the encroachment of the deserts in Africa helped weaken Roman civilization and eventually contributed to the collapse of the empire between A.D. 400 and 500.

In the Americas we see a similar pattern at work. The Mayan civilization flourished in what is now the southern part of Mexico and northern Central America. Although Mayan society and population grew slowly beginning around 1200 B.C., by A.D. 800, the major city of Tikal had fifty thousand inhabitants. Cities in the region had similar architecture and contained ceremonial centers with steep pyramid temples that were over one hundred feet high. The Mayans also developed a complex and highly accurate calendar. Despite this sophistication, within a few decades after A.D. 800, the society began to disintegrate. Today, only a small number of peasants live in the area.

The primary cause of the collapse of Mayan civilization was a lack of food, which resulted from unsustainable demands that were placed on the land. Fragile soils of the tropical rain forest are easily eroded once exposed. Once the land was cleared for agriculture, soil erosion increased and crop yields eventually declined. When the land became exhausted, the surplus food disappeared, which led to conflict between the peasants who grew the food and the ruling elite and their armies in the cities. The outcome was a catastrophic fall in population. The Mayans were unable to maintain the elaborate culture they had built on their fragile environmental base. Civilization collapsed rapidly.[49] Within a few decades the great cities were abandoned, to be reclaimed by the jungles, where they remained hidden from the larger world for the next thousand years.

The main lesson that I draw from these stories is that we need to take modern-day adversity trends seriously. History is filled with too many examples of civilizations that made excessive demands on the environment and then disintegrated. In each case, the pattern is similar. The rise of a civilization begins with the development of intensive agriculture, which generates a food surplus. Unsustainable agricultural practices eventually produce deforestation and soil erosion, with a subsequent loss in agricultural productivity and the collapse of the civilization. Does the human family need to learn lessons at a global level that we have already learned many times before at a smaller scale? Or will we wake up together to the lessons of history and choose a sustainable pathway into the future?

Chapter Three

A New Perceptual Paradigm: We Live in a Living Universe

In the history of the collective as in the history of the individual, everything depends on the development of consciousness.
—*Carl Jung*

The whole of life lies in the verb *seeing.*
—*Teilhard de Chardin*

Humanity's Fourth Awakening

IN the previous chapter, we considered powerful adversity trends that are pushing humanity toward an evolutionary crash. In the next four chapters, we turn to consider equally powerful opportunity trends that are pulling us toward a positive future—an evolutionary bounce.

The first opportunity trend that could transform our impending crash into a spectacular bounce is a shift in our shared view of the universe—from thinking of it as dead to experiencing it as alive. In regarding the universe as alive and ourselves as continuously sustained within that aliveness, we see that we are intimately related to everything that exists. This startling insight—that we are cousins to everything that exists in a living, continuously regenerated universe—represents a new way of looking at and relating to the world and overcomes the profound separation that has marked our past. From the combined wisdom of science and spirituality is emerging an understanding that could provide the perceptual foundation for the diverse people of the world to come together in the shared enterprise of building a sustainable and meaningful future.

For some, a shift in perception may seem so subtle as to be inconsequential. Yet, all of the deep and lasting revolutions in human development have been generated from just such shifts. Only three times before in human experience has our view of reality been so thoroughly transformed that it has created a revolution in our sense of ourselves, our relationships with others, and our view of the universe. I want to mention these briefly here and then explore them more fully later (in chapter eight, "Humanity's Central Project").

The first transformation in our view of reality and identity occurred when humanity "awakened" roughly thirty-five thousand years ago. The archeological record shows that the beginnings of a reflective consciousness emerged decisively at this time as numerous developments were occurring in stone tools, burial sites, cave art, and migration patterns. Because we were just awakening to our capacity for "knowing that we know," we were surrounded by mystery at every turn. Nonetheless, human culture was born in these first glimmerings of personal and shared awareness.

The second time our view of reality and human identity changed dramatically was roughly ten thousand years ago, when humanity shifted from a nomadic life to a more settled existence in villages and farms. It was midway during the agrarian period, roughly five thousand years ago, that we see the rise of city-states and the beginnings of civilization.

The third time that our perceptual paradigm was transformed was roughly three hundred years ago, when the stability of agrarian society gave way to the radical dynamism and materialism of the scientific-industrial era. Each time that humanity's prevailing paradigm has changed, all aspects of life have changed with it, including the work that people do, the ways they live together, how they relate to one another, and how they see their role in society and place in the universe.

A paradigm is our way of looking at and thinking about ourselves and everything around us. It is the frame of mind out of which we operate. Our paradigm sets the limits on what thoughts we can think, what emotions we can feel, and what reality we can perceive. Willis Harman, renowned futurist, described a paradigm as "the basic way of perceiving, thinking, valuing, and doing associated with a particular vision of reality."[1] A paradigm tells most people, most of the time, what's real and what's not, what's important and what's not, and how things are related to one another. A paradigm is more than a dry mental map—it is our window onto the world that shapes how we see and understand the nature of reality, our sense of self, and our feelings of social connection and purpose.

We are now living at a time when humanity's perceptual paradigm is undergoing one of its rare shifts, and that shift has the potential to dramatically transform life for each of us. A paradigm shift therefore goes to the core of people's lives. It is much more than a change in ideas and how we think. It is a change in our view of reality, identity, social relationships, and human purpose. A paradigm shift can be felt in the body, heart, mind, and soul.

How do paradigm shifts occur? Paradigms operate beneath the surface of popular culture, largely unnoticed until the old way of perceiving begins to generate more problems than it solves. These problems then become the catalyst for triggering the shift to the next paradigm, which opens up new opportunities. When we first enter a new civilizational paradigm (such as during the shift from the agricultural era to the industrial era), we experience new freedoms and creative potentials. As we fulfill the potentials of that paradigm, however, it eventually becomes a constricting framework. Its partial or incomplete nature leads to a crisis, which in turn leads to a breakthrough into the next, more spacious paradigm, in which a new level of learning and creative expression can unfold.

The paradigm of the scientific-industrial era, while it has afforded great benefits, is now generating far more problems than it is solving. These problems are catalysts for a paradigm shift. They are forcing us to expand our perceptual horizons to a higher and more inclusive level. Albert Einstein described a paradigm shift by saying that we cannot solve problems at the same level at which they are created.

The emerging paradigm represents a convergence of insights from modern science and the world's spiritual traditions. At the heart of the new paradigm is a startling idea—that our cosmos is not a fragmented and lifeless machine (as we have believed for centuries) but is instead a unified and living organism. Although it is new for our times, the idea that the universe is alive is an ancient one. More than two thousand years ago, Plato described the universe as "one Whole of wholes" and "a single Living Creature which encompasses all of the living creatures that are within it."[2] What is unprecedented is how this notion is being informed today by both modern science and the world's diverse spiritual traditions. Let us look at the evidence from both these sources, beginning with recent scientific discoveries. We shall explore then the implications of this paradigm shift, and the opportunity it presents for imagining and building a sustainable future.

Scientific Evidence of a Living Universe

LESS than a hundred years ago, when Einstein was developing his theory of relativity, he considered the universe a static, unchanging system no larger than the cloud of stars that we now know to be our galaxy. Today, we know that the universe is expanding rapidly and contains at least 50 billion galaxies, each with a 100 billion or more stars. What is more, we now know that our cosmos embodies an exquisitely precise design. Researchers have calculated that if the universe had expanded ever so slightly faster or slower than it did (even by as little as a trillionth of a percent), the matter in our cosmos would have either quickly collapsed back into a black hole or spread out so rapidly that it would have evaporated. Such amazing precision implies a living intelligence is at work. Beyond these surprising findings are even more extraordinary conclusions that, taken together, suggest our universe is a living system.

Although there is no clear agreement among scientists as to what constitutes a living system, it seems reasonable that if our cosmos is alive, it would exhibit specific properties that are characteristic of all life—such as being a unified entity, having some form of consciousness, and being able to reproduce itself. As we shall explore, these are among the properties of our universe that are emerging from modern science.

The cosmos is a unified system. Physicists used to view our universe as being composed of separate fragments. Today, however, despite its unimaginably vast size, the universe is increasingly regarded as a single functioning system. Because other galaxies are millions of light-years away, they appear so remote in space and time as to be separate from our own. Yet, scientific experiments show that things that seem to be separate are actually connected in fundamental ways that transcend the limitations of ordinary space and time.[3] Described as "nonlocality," this is one of the most stunning insights from modern science. Even though we live in a world of apparent separation, the new physics describes the more fundamental reality as that of seamless interconnection. Physicist David Bohm says that ultimately we have to understand the entire universe as "a single undivided whole."[4] Instead of separating the universe into living and nonliving things, Bohm sees animate and inanimate matter as inseparably interwoven with the life-force that is present throughout the universe and includes not only matter but also energy and seemingly empty space. For Bohm, then, even a rock has its unique form of aliveness. Life is dynamically flowing through the fabric of the entire universe.[5]

Our home galaxy—the Milky Way—is a swirling, disk-shaped cloud containing 100 billion or so stars. It is part of a local group of nineteen galaxies (each with 100 billion stars), which in turn is part of a larger Local Supercluster of thousands of galaxies. This supercluster resembles a giant, many-petaled flower. Beyond this, astronomers estimate that there are perhaps 100 billion galaxies in the observable universe (each with 100 billion or so stars). Scientists and spiritual seekers alike ask the question: If this is a unified system, then could all this be but a single cell within a much greater organism?[6]

It contains immense amounts of background energy. In the new view of reality, an extraordinary amount of energy permeates the cosmos. Empty space is not actually empty. Even in a complete vacuum, there exist phenomenal levels of background energy called "zero point energy." Bohm calculated that a single cubic centimeter of "empty space" contained the energy equivalent of millions of atomic bombs.[7] This is not simply a theoretical abstraction. A number of people are working to create energy devices that can tap into this background energy. Our universe is permeated by and exists within a vast ocean of flowing life energy.

The cosmos is continuously regenerated. For decades, the dominant cosmology in contemporary physics has held that creation ended with the Big Bang some 13 billion years ago and that since then nothing more has happened than a rearranging of the cosmic furniture. Because traditional physicists think of creation as a one-time miracle from "nothing," they regard the contents of the universe—such as trees, rocks, and people—as being constituted from ancient matter. In sum, the dead-universe theory assumes creation occurred billions of years ago, when a massive explosion spewed out lifeless material debris into equally lifeless space and has, by random processes, organized itself into life-forms on the remote planet-island called Earth.

In striking contrast, the living-universe theory describes the cosmos as a unified system that is completely re-created at each moment. Unlike traditional physicists who believe that creation ended with the miraculous birth of the cosmos billions of years ago, living-universe theorists hold that the cosmos continues to be maintained, moment by moment, by an unbroken flow-through of energy. They compare the cosmos to the vortex of a tornado or a whirlpool—as a completely dynamic structure. David Bohm calls the universe an "undivided wholeness in flowing movement."[8] In this view, our universe has no freestanding material existence of its own. The notion of continuous creation is even more remarkable when we consider that it includes not only matter but also the fabric of seemingly "empty" space. Space is not a simple emptiness waiting to be filled, but is itself a dynamically constructed transparency. Therefore, the entire cosmos is being regenerated at each instant in a single symphony of expression that unfolds from the most minute aspects of the subatomic realm to the vast reaches of thousands of billions of galactic systems. The whole cosmos, all at once, is the basic unit of creation.

It utterly overwhelms the imagination to consider the size and complexity of our cosmos with its billions of galaxies and trillions of planetary systems, all partaking in a continuous flow of creation. How can it be so vast, so subtle, so precise, and so powerful? Metaphorically, we inhabit a cosmos whose visible body is billions of light years across, whose organs include billions of galaxies, whose cells include trillions of suns and planetary systems, and whose molecules include an unutterably vast number and diversity of life-forms. The entirety of this great body of being, including the fabric of space-time, is being continuously regenerated at each instant. Scientists sound like poets as they attempt to describe our cosmos in its process of becoming. The mathematician Norbert Wiener expresses it this way: "We are not stuff that abides, but patterns that perpetuate themselves; whirlpools of water in an ever-flowing river."[9] Physicist Max Born writes, "We have sought for firm ground and found none. The deeper we penetrate, the more restless becomes the universe; all is rushing about and vibrating in a wild dance."[10] Physicist Brian Swimme tells us, "The universe emerges out of an all-nourishing abyss not only 13 billion years ago but in every moment."[11]

The new physics allows us to see everything in the cosmos as a flowing movement that co-arises along with everything else, moment-by-moment, in a process of continuous regeneration. If all is in motion at every level, and all motion presents itself as a coherent and stable pattern, then all that exists is profoundly orchestrated. All flows comprise one grand symphony in which we are all players, a single creative expression—a uni-verse.

Freedom is at its foundations. Another shift in the scientific view of the universe has to do with views about the existence of freedom. Whereas traditional physicists have seen the cosmos as being like a clockwork mechanism that is locked into predetermined patterns of development, the new physics sees it as a living organism that has the freedom and spontaneity to grow in unexpected ways. Freedom is at the very foundation of our cosmos. Uncertainty (and thus freedom) is so fundamental that quantum physics describes reality in terms of probabilities, not certainties. No one part of the cosmos determines the functioning of the whole; rather, everything seems to be connected with everything else, weaving the cosmos into one vast interacting system. Everything that exists contributes to the cosmic web of life at each moment, whether it is conscious of its contribution or not. In turn, it is the consistency of interrelations of all the parts of the universe that determines the condition of the whole. We therefore have great freedom to act within the limits established by the larger web of life within which we are immersed.

A living universe is a learning system in which we are free to make mistakes and to change our minds. In other words, if the universe is being continuously re-created, then each moment provides an opportunity for a fresh start. This is how the philosopher Renée Weber describes the creative and experimental nature of the universe: "Through us, the universe questions itself and tries out various answers on itself in an effort—parallel to our own—to decipher its own being."[12] Every moment, the universe re-creates itself and provides us with an opportunity to exercise our basic freedom to do the same.

Consciousness is present throughout. Consciousness, or a capacity for self-organizing reflection or knowing, is basic to life. If the universe is alive, we should therefore find evidence of some form of consciousness operating at every level—and that is exactly what we find. The respected physicist Freeman Dyson writes this about consciousness at the quantum level: "Matter in quantum mechanics is not an inert substance but an active agent, constantly making choices between alternative possibilities . . . It appears that mind, as manifested by the capacity to make choices, is to some extent inherent in every electron."[13] This does *not* mean that an atom has the same consciousness as a human being, but rather that an atom has a reflective capacity appropriate to its form and function.

Consciousness is present even at the primitive level of molecules consisting of no more than a few simple proteins. Researchers have found that such molecules have the capacity for complex interaction that is the signature of living systems. As one of the researchers who made this discovery stated, "We were surprised that such simple proteins can act as if they had a mind of their own."[14]

At a somewhat higher level, we find consciousness operating in the remarkable behavior of a forest slime mold in search of a new feeding area. For most of its life, slime mold exists as a single-cell amoeba. When it needs food, however, it can transform itself into a much larger entity with new capacities. Individual amoebas send out signals to others nearby until thousands assemble. When they reach a critical mass, they organize themselves, without the aid of any apparent leader, into an organism that can move across the forest floor. Upon reaching a better feeding area, they release spores from which new amoebas are formed.[15] Thus, under conditions of great stress, the forest slime mold is able to mobilize a capacity for collective consciousness and action so as to ensure its own survival.

If some form of consciousness is operating at the level of atoms, molecules, and single-cell organisms, we should not be surprised to find that consciousness is a basic property of the universe that is manifest at every level. Scientific investigation of intuitive or psychic abilities in humans provides further insight into the nature and ecology of consciousness. Dean

Radin, director of the Consciousness Research Laboratory at the University of Nevada, did an exhaustive analysis of parapsychological or psi research involving more than eight hundred studies and sixty investigators over nearly three decades. Based on this research, he concluded that consciousness includes both "receiving" and "sending" potentials.

Evidence of the receiving potentials of consciousness comes from experiments concerned with perception at a distance, which is sometimes called "remote viewing." This is the ability to receive meaningful information by nonphysical means about a remote person or location simply by opening our knowing faculty to that possibility. In remote viewing, the receiver does not acquire exact information but rather intuitive impressions regarding, for example, where a person might be located or his state of well-being. Radin found that remote viewing has "been repeatedly observed by dozens of investigators using different methods."[16] He concluded that a capacity for conscious knowing "operates between minds and through space."

Evidence of the sending potentials of consciousness comes from experiments dealing with mind-matter interactions, such as influencing the swing of a pendulum clock. Radin concluded that "after sixty years of experiments . . . researchers have produced persuasive, consistent, replicated evidence that mental intention is associated with the behavior of physical systems."[17]

I would have been reluctant to write about consciousness being a basic property of the universe—and in particular about parapsychology—had I not had an unusual opportunity to learn about it firsthand. During the early 1970s, I worked as a senior social scientist at the Stanford Research Institute, a large think-tank south of San Francisco now called SRI International. There I did studies of the long-range future, primarily for government agencies such as the President's Science Advisor and the Environmental Protection Agency. While doing this work, I was invited to participate in parapsychological experiments being conducted at SRI by two senior physicists, Dr. Hal Puthoff and Dr. Russell Targ. Several days a week for three years I went to their laboratory to take part in both formal and informal experiments.

One series of formal experiments involved remote viewing. The procedure was simple. I would be locked in a bare room with a pad of paper, a pencil, and a tape recorder and asked to describe where in the Bay Area one of the experimenters would be. After my door was locked, his destination was selected from a pool of more than a hundred possible locations by drawing an envelope at random from a locked safe. My task, after waiting a half hour for him to travel to his destination, was to describe in words or drawings the location of this outbound person. Was he in a boat on the bay? In a car on the freeway? In a grove of redwood trees? In a movie theater? In the room next door? My only instructions were, "Take a deep breath, close your eyes, and tell us what you see." Although my impressions were subtle and fleeting, I gradually learned that we all have an intuitive ability to "see" at a distance. Through our intuition, each of us can acquire useful impressions, images, and insights about a person or place that is distant from us. In my experience and that of the other subjects, the drawings and descriptions were often sufficiently accurate to allow independent judges to differentiate significantly among the various targets and, at levels far beyond chance, to match many of those descriptions closely with the actual locations.[18]

Another series of experiments involved working with a

computer that would randomly select one of four buttons prominently displayed on top of it. My task was to intuitively discover which of the four had been selected and to press the correct button. More than seven thousand selections were tallied under controlled conditions—an exhausting process requiring intense concentration over dozens of test sessions. The overall results were significantly above chance.[19]

These grueling experiments convinced me that we do have an intuitive connection with the universe; they also demonstrated that our capacity to use our intuition is still in its infancy given our early stage of learning. The most important insight that I take away from these and other experiments is that we *all* have an intuitive faculty. An empathic connection with the universe is nothing special; it is built into the workings of the cosmos. Participating in these experiments showed me that our being does not stop at the edge of our skin but extends into and is inseparable from the universe.

If consciousness is found at every level of the cosmos and, further, is not confined within the brain, but extends beyond the body and can meaningfully interact with the rest of the universe in both sending and receiving communications, then this is striking evidence that our cosmos is subtly sentient, responsive, conscious—and alive. The physicist Freeman Dyson thinks it is reasonable to believe in the existence of a "mental component of the universe." He says, "If we believe in this mental component of the universe, then we can say that we are small pieces of God's mental apparatus."[20] While it is stunning to consider that every level of the cosmos has some degree of consciousness, that seems no more extraordinary than the widely accepted view among scientists that the cosmos emerged as a pinpoint some 13 billion years ago as a "vacuum fluctuation" where nothing pushed on nothing to create everything.

The cosmos seems able to reproduce itself. A key attribute of any living system is its ability to reproduce itself. A startling finding from the new physics is that our cosmos may very well be able to reproduce itself through the functioning of black holes. In his book, *In the Beginning: The Birth of the Living Universe,* astrophysicist John Gribbin explains that the explosion of our universe in the Big Bang is the time-reversed mirror image of the collapse of a massive object into a black hole. Many of the black holes that form in our universe, he reasons, may thus represent the seeds of new universes: "Instead of a black hole representing a one-way journey to nowhere, many researchers now believe that it is a one-way journey to somewhere—to a new expanding universe in its own set of dimensions."[21] Gribbin's dramatic conclusion is that "our own universe may have been born in this way out of a black hole in another universe." He explains it in this way:

> If one universe exists, then it seems that there must be many—very many, perhaps even an infinite number of universes. Our universe has to be seen as just one component of a vast array of universes, a self-reproducing system connected only by the "tunnels" through spacetime (perhaps better regarded as cosmic umbilical cords) that join a "baby" universe to its "parent."[22]

Gribbin suggests not only that universes are alive, but also that they evolve as other living systems do: "Universes that are 'successful' are the ones that leave the most offspring."[23]

The idea of many universes evolving through time is not new. David Hume noted in 1779 that many prior universes "might have been botched and bungled throughout an eternity ere this system."[24]

Is the cosmos a living system? It certainly appears so in the light of recent scientific findings. Our universe is revealing itself to be a profoundly unified system in which the interrelations of all the parts, moment-by-moment, determine the condition of the whole. Our universe is infused with an immense amount of energy, and is being continuously regenerated in its entirety, while making use of a reflective capacity or consciousness throughout. As an evolving, growing, and learning system, it is natural that freedom exists at the quantum foundations of the universe. It even appears that the universe has the ability to reproduce itself through the vehicle of black holes. When we put all of these properties together, it suggests an even more spacious view of our cosmic system. Our universe is a living system of elegant design that was born from and is continuously regenerated within an even larger universe. *We are living within a "daughter universe" that, for 13 billion years, has been living and growing within the spaciousness of a Mother Universe.* The Mother Universe has existed forever, holding countless daughter universes in its grand embrace while they grow and mature through an eternity of time.

The Mother Universe

WHEN our cosmos blossomed into existence from an area smaller than a pinpoint some 13 billion years ago, it emerged out of "somewhere." Modern physics is beginning to speculate on the nature of this generative ground. The distinguished Princeton astrophysicist John Wheeler describes space as the basic building block of reality. He explains that material things are "composed of nothing but space itself, pure fluctuating space . . . that is changing, dynamic, altering from moment to moment." Wheeler goes on to say that "of course, what space itself is built out of is the next question . . . The stage on which the space of the universe moves is certainly not space itself . . . The arena must be larger: *superspace* . . . [which is endowed] with an infinite number of dimensions."[25] What Wheeler calls "superspace," I am calling the Mother Universe.

The idea of a "superspace" or Mother Universe is not simply a creation of theoretical physics. It is a reality that can be directly experienced and has ancient roots in the world's meditative traditions. For example, more than twenty centuries ago, the Taoist sage Lao-tzu, described it this way:

> There was something formless and perfect
> before the universe was born.
> It is serene. Empty.
> Solitary. Unchanging.
> Infinite. Eternally present.
> It is the mother of the universe.
> For lack of a better name,
> I call it the Tao.[26]

Regardless of what the Mother Universe is called, all wisdom traditions agree that it is ultimately beyond description. Nevertheless, many attempts have been made to describe her paradoxical qualities. Here are six of the key attributes of the Mother Universe as seen by both East and West:

- **Present everywhere**—The clear, unbounded life-energy of the Mother Universe is present in all material forms as well as in seemingly empty space. The Mother Universe is not separate from us, nor is it other than the "ordinary" reality that is continuously present around us. The Mother Universe is also not limited to containing only our universe; there likely are a vast number of other universes growing in other dimensions of her unimaginable spaciousness.
- **Nonobstructing**—The Mother Universe is a living presence out of which all things emerge, but it is not itself filled or limited by these things. Not only are all things in it; it is in all things. There is mutual interpenetration without obstruction.
- **Utterly impartial**—The Mother Universe allows all things to be exactly what they are without interference. We have immense freedom to create either suffering or joy.
- **Ultimately ungraspable**—The power and reach of the Mother Universe is so vast that it cannot be grasped by our thinking mind. As the source of our existence, the Mother Universe is forever beyond the ability of our limited mental faculties to capture conceptually.
- **Compassionate**—Boundless compassion is its essence. To experience the subtle and refined resonance of the Mother Universe is to experience unconditional love.

- **Profoundly creative—**Because we humans do not know how to create a single flower or cubic inch of space, the creative power of the Mother Universe to bring into existence and sustain entire cosmic systems is utterly incomprehensible.

It is useful to contemplate these extraordinary characteristics of the Mother Universe so as to awaken ourselves to the profound miracle in which we are immersed. In that spirit, here is an evocative portion of what the Chinese monk Shao has written in describing what I call the Mother Universe:

> If you say that It is small,
> It embraces the entire universe.
> If you say It is large,
> It penetrates the realm of atoms.
> Call It one; It bears all qualities.
> Call It many; Its body is all void.
> Say It arises; It has no body and no form.
> Say It becomes extinct; It glows for all eternity.
> Call It empty; It has thousands of functions.
> Say It exists; It is silent without shape.
> Call It high; It is level without form.
> Call It low; nothing is equal to It.[27]

In looking across the world's spiritual traditions, the insight emerges again and again: although we live in a world of seeming separation and division, our universe is a unified whole brimming with life and infused with a divine presence. Here are a few examples:

"Earth's crammed with Heaven, and every common bush afire with God."

—*Elizabeth Barrett Browning,* poet

The Tao is the sustaining life-force and the mother of all things; from it, all "things rise and fall without cease."[28]

—Taoist tradition

"Heaven and earth and I are of the same root . . . are of one substance."[29]

—*Sojo,* a Zen monk

Jesus was asked, "When will the kingdom come?" He replied, "It will not come by waiting for it. . . . Rather, the Kingdom of the Father is spread out upon the earth, and men do not see it."[30]

—*Gospel of Thomas,* Gnostic Gospels

"For those who are awake the cosmos is one."[31]

—*Heraclitus,* ancient Greek philosopher

"My solemn proclamation is that a new universe is created every moment."[32]

—*D. T. Suzuki,* Zen scholar and teacher

"I am in some sense boundless, my being encompassing the farthest limits of the universe, touching and moving every atom of existence. The same is true of everything else . . . It is not just that 'we are all in it' together. We all *are* it, rising and falling as one living body."[33]

—*Francis Cook,* Buddhist scholar,
describing Hua-yen Buddhism

"All Hindu religious thought denies that the world of nature stands on its own feet. It is grounded in God; if he were removed it would collapse into nothingness."[34]

—*Huston Smith,*
scholar of the world's sacred traditions

"There is a life pouring into the world, and it pours from an inexhaustible source."[35]

—*Joseph Campbell,*
scholar of world's creation stories

"Creation, then, is an ongoing story of new beginnings, opportunities to begin again and again. God began to create, is still creating; nothing is finished."[36]

—*Wayne Muller,* ordained minister

"God is creating the entire universe, fully and totally, in this present now. Everything God created . . . God creates now all at once."[37]

—*Meister Eckhart,* Christian mystic

"The entire cosmos comes forth moment by moment from this one fundamental innate mind of clear light."[38]

—*Lex Hixon,*
scholar of the world's sacred traditions

Christians, Buddhists, Hindus, Jews, Muslims, Taoists, mystics, tribal cultures, and Greek philosophers have all given remarkably similar descriptions of the universe and the life-force that pervades it. These are more than poetic and metaphorical descriptions. Because we find the notion of a living universe emerging across cultures and millennia as well as from modern science, there is compelling evidence that it forms the basis of a powerful perceptual paradigm—one that will open up enormous opportunities for the human family as we are pressed to create a sustainable future for ourselves.

Implications of the Living Universe Paradigm

LIKE any paradigm shift, the shift to a living universe paradigm is transformative. In addition to changing our view of the universe, it can alter our sense of identity, our sense of life purpose, how we relate with others, and much more. Let's consider a few of its many implications.

A rebirth of connectedness in all aspects of life. To explore how our experience of the world might change with a shift to a living universe paradigm, let's look at how American Indians perceived and experienced the world. Their culture provides a clear window into the experience of living with an infusing aliveness that is an intimate part of everyday life.

Author Luther Standing Bear expresses the wisdom of indigenous peoples around the world when he says that for the Lakota Sioux, "there was no such thing as emptiness in the world. Even in the sky there were no vacant places. Everywhere there was life, visible and invisible, and every object gave us a great interest in life. The world teemed with life and wisdom; there was no complete solitude for the Lakota."[39] For the Lakota, who inhabited the upper Midwest of the United States, religion was based on a direct experience of an all-pervading spirit throughout the world. Since a life-force was felt to be in and through everything, all things were seen as being connected and related. Because everything is an expression of the Great Spirit, everything deserves to be treated with respect.

This paradigm was not unique to the Lakota. One of the most dense concentrations of Indian populations in North America—the Ohlones—lived along the fertile region that is now San Francisco, Oakland, San Jose, and Monterey in California.[40] The Ohlones lived sustainably on this land for four thousand to five thousand years. Like the Lakota, their religion was without dogma, churches, or priests because it was so pervasive, like the air. Malcolm Margolin describes their experience of the world in his book, *The Ohlone Way*:

> The Ohlones, then, lived in a world perhaps somewhat like a Van Gogh painting, shimmering and alive with movement and energy in ever-changing patterns. It was a world in which thousands of living, feeling, magical things, all operating on dream-logic, carried out their individual actions . . . Power was everywhere, in everything, and therefore every act was religious. Hunting a deer, walking on a trail, making a basket, or pounding acorns were all done with continual reference to the world of power.[41]

In shifting to the living universe paradigm, we rediscover the aliveness that is at the foundation of the universe, and we realize that we are not disconnected from the larger universe, and never have been. An Ojibwe Indian poem expresses this realization beautifully:

> Sometimes I go about pitying myself,
> and all the while I am being carried
> on great winds across the sky.

With a cosmology of a living universe, a shining miracle exists everywhere. There are no empty places in the world.

Everywhere there is life, both visible and invisible. All of reality is infused with wisdom and a powerful presence.

The awakening of cosmic identity. In the industrial era paradigm, we are no more than biological beings, ultimately separate from others and the rest of the universe. The new findings from physics, however, reveal that we are intimately connected with the entire cosmos. Our actual identity or experience of who we are is vastly bigger than we thought—we are moving from a strictly personal consciousness to a conscious appreciation of ourselves as integral to the cosmos. Physicist Brian Swimme explains that the intimate sense of self-awareness we experience bubbling up at each moment "is rooted in the originating activity of the universe. We are all of us arising together at the center of the cosmos."[42] We thought that we were no bigger than our physical bodies, but we are discovering that we are beings of cosmic dimension, part of the flow of continuous re-creation of the cosmos. By becoming aware of that stream of life in our direct experience, we become conscious of our connection with the living universe.

Technically, we humans are more than *Homo sapiens* or "wise"—we are *Homo sapiens sapiens* or "doubly wise."[43] In other words, whereas animals "know," humans have the capacity to "know that we know." In the new paradigm, our sense of identity takes on a paradoxical and mysterious quality: we are both observer and observed, knower and that which is known. We are each completely unique yet completely connected with the entire universe. There will never be another person like any one of us in all eternity, so we are absolutely original beings. At the same time, since our existence arises from and is woven into the deep ecology of the universe, we are completely integrated with all that exists. Awakening to the miraculous nature of our identity as simultaneously unique and interconnected with a living universe can help us overcome the species arrogance and sense of separation that threaten our future.

Living lightly in a living universe. In a dead universe, materialism makes sense; in a living universe, simplicity makes sense. Let's consider these two alternatives.

If the universe is unconscious and dead at its foundations, then each of us is the product of blind chance among materialistic forces. It is only fitting that we the living exploit on our own behalf that which is not alive. If the universe is lifeless, it has no larger purpose or meaning, and neither does human existence. If we are separate beings in a lifeless universe, there are no deeper ethical or moral consequences to our actions beyond their immediate, physical impacts. It is only natural, therefore, that we focus on consuming material things to minimize life's pains and maximize its comforts.

On the other hand, if the universe is conscious and alive, then we are the product of a deep-design intelligence that infuses the entire cosmos. We shift from feelings of existential isolation in a lifeless universe to a sense of intimate communion within a living universe. If life is nested within life, then it is only fitting that we treat everything that exists as alive and worthy of respect. Our sense of meaningful connection expands to the entire community of life, including past, present, and future generations. Every action in a living universe is felt to have ethical consequences as it reverberates throughout the ecosystem of the living cosmos. The focus of life shifts from a desire for high-consumption lifestyles (intended to provide both material pleasures and protection from an indifferent universe) toward sustainable and simple ways of living (intended to connect us with a purposeful universe of which we are an integral part). In a living universe, it is only natural that people would choose simpler ways of living that afford greater time and opportunity for meaningful relationships, creative expression, and rewarding experiences. As we consciously explore our connection with a living universe, concern with material consumption will naturally tend to shift into the background of our lives.

Living with purpose in a living universe. The shift to a new paradigm also brings a shift in our sense of evolutionary purpose. We are shifting from seeing our journey as a secular adventure in a fragmented and lifeless cosmos without apparent meaning or purpose, to seeing it as a sacred journey through a living and unified cosmos. Our primary purpose is to embrace and learn from both the pleasure and the pain of the world. If there were no freedom to make mistakes, there would be no pain. If there were no freedom for authentic discovery, there would be no ecstasy. In freedom, we experience both pleasure and pain in the process of discovering our identity as beings of both earthly and cosmic dimensions. In the words of the Australian aborigines, we are learning how to survive in infinity.

Living ethically in a living universe. A form of natural ethics accompanies our intuitive connection with a living universe. When we are truly centered in the life current flowing through us, we tend to act in ways that promote the well-being and harmony of the whole. Our connection with the Mother Universe provides us with a sort of moral tuning fork that makes it possible for individuals to come into collective alignment. An underlying field of consciousness weaves humanity together, making it possible for us to understand intuitively what is healthy and what is not, what works and what doesn't. We can each tune into this living field and sense what is in harmony with the well-being of the whole. When we are in alignment, we experience—as a sort of kinesthetic sense—a positive hum of well-being.[44] In a similar way, we also experience the hum of discordance.

The new paradigm will usher us into an era in which people will be inclined to live ethically because they understand that everything they do is woven into the infinite depths of the Mother Universe. In his *Book of Mirdad,* Mikhail Nimay describes this insight beautifully:

> So think as if your every thought were to be etched in fire upon the sky for all and everything to see. For so, in truth, it is.
> So speak as if the world entire were but a single ear intent on hearing what you say. And so, in truth, it is.
> So do as if your every deed were to recoil upon your head. And so, in truth, it is.
> So wish as if you were the wish. And so, in truth, you are.[45]

When we discover that all beings are part of the seamless fabric of creation, it naturally awakens in us a sense of connection with and compassion for the rest of life. We automatically broaden our scope of empathy and concern when we realize that we are inseparable from all that exists. We no longer see ourselves as isolated entities whose being stops at the edge of our skin, and whose empathy stops with our family, or our race, or our nation. We see that, because we all arise simultaneously from a deep ocean of life-energy, a vital connection exists among all beings.

The emergence of a living universe paradigm is not simply a lateral shift from one set of values to another; it is a contextual shift, from one cultural atmosphere to another, from one perceptual environment to another. It transforms the human story. After 13 billion years of evolution, we stand upon the Earth as agents of self-reflective and creative action on behalf of the universe. We see that we are participants in an unceasing miracle of creation. This recognition brings a new confidence that our potentials are as exalted, magnificent, and mysterious as the living universe that surrounds and sustains us.

Chapter Four

Choosing a New Lifeway: Voluntary Simplicity

The price of anything is
the amount of life that you have to pay for it.
—*Henry Thoreau*

Too many people spend money they haven't earned,
to buy things they don't want,
to impress people they don't like.
—*Will Rogers*

A Quiet Revolution

THE second opportunity trend that can make an enormous contribution to an evolutionary bounce is a voluntary shift toward more sustainable and satisfying ways of living. This is a promising development for, in order to meet the coming evolutionary challenges successfully, I believe that we will need to make major changes in every aspect of our lives—including the transportation we use, the food that we eat, the homes and communities that we live in, the work that we do, and the education that we provide. Although it is appealing to think that marginal measures such as intensified recycling and more fuel-efficient cars will take care of things, they will not. We need to make sweeping changes—both externally and within ourselves. A sustainable future will demand far more than a surface change to a different *style* of life—it requires a deep change to a new *way* of life.

Is it realistic to think that a new way of life could emerge? The American Dream is founded on the premise that the more you consume, the happier and more satisfied you will be. But decades of social science research reveal that, except for the very poor, our level of income has no significant effect on our level of satisfaction with life. As soon as we reach a comfortable level of income, the correlation between income and happiness diminishes dramatically.[1] Studies of entire nations reveal a similar pattern. For example, in the United States, while per-capita disposable income (adjusted for inflation) doubled between 1960 and 1990, the percentage of Americans reporting they were "very happy" remained essentially the same (35 percent in 1957 and 32 percent in 1993).[2]

In an article in *The New York Times* on the high price of the pursuit of affluence, Alfie Kohn says that researchers have amassed significant evidence that "satisfaction simply is not for sale."[3] In fact, Kohn says that "people for whom affluence is a priority in life tend to experience an unusual degree of anxiety and depression as well as a lower overall level of well-being." The single-minded pursuit of affluence actually reduces people's sense of well-being and satisfaction. This is the dark side of the dream of getting rich, and it seems to hold true regardless of age, level of income, or culture. Researchers have also found that "pursuing goals that reflect genuine human needs, like wanting to feel connected to others, turns out to be more psychologically beneficial than spending one's life trying to impress others." Lily Tomlin seems to be right when she says, "the trouble with being in the rat race is that even if you win, you're still a rat."

Are people waking up to another way of life, focused not on the pursuit of affluence, but on close and caring relationships, a rich inner life, and creative contributions to the world? Is there a new way of life emerging that pulls back from materialism not out of sacrifice but in an attempt to find authentic and lasting sources of satisfaction and meaning?

Amid a frenzy of conspicuous consumption, an inconspicuous revolution has been stirring. A growing number of people are seeking a way of life that is more satisfying and sustainable. This quiet revolution is being called by many names; including voluntary simplicity, soulful simplicity, and compassionate living. But whatever its name, its hallmark is a new common sense—namely, that life is too deep and consumerism is too shallow to provide soulful satisfaction. As a result, more and more people, particularly in the United States and Europe, have been exploring life beyond advertising's lure. These people have experienced the good life that consumerism has to offer and found it flat and unsatisfying compared to the rewards of the simple life. Their choice of a lifeway of conscious simplicity is driven not by sacrifice but by a growing understanding of the real sources of satisfaction and meaning—gratifying friendships, a fulfilling family life, spiritual growth, and opportunities for creative learning and expression.

This is a leaderless revolution—a self-organizing movement where people are consciously taking charge of their lives. It is a clear and promising example of people growing up and taking responsibility for how their lives connect with the Earth and the future. Many of these lifeway pioneers have been working at the grassroots level for several decades, often feeling alone, not realizing that scattered through society are others like themselves numbering in the millions.

What Is Voluntary Simplicity?

THERE has been a tendency in the mainstream media to equate a simple way of life with a lifestyle of material frugality and then to focus on the material changes people are making, such as recycling, buying used clothing, and planting gardens. While these are a few of the visible expressions of the simple life, this portrayal misses much of the juice, joy, and purpose of simple living. The overwhelming majority of those choosing a life of simplicity are not seeking to fulfill some romantic notion of returning to nature. Instead, they are seeking greater sanity and soulfulness in a society in which separation from nature is rampant. For the most part, these lifeway pioneers are not moving back to the land; they are making the most of wherever they are by crafting a way of life that is more satisfying and sustainable.

Richard Gregg, my mentor on the subject of simplicity, wrote in 1936 that the purpose of life was, fundamentally, to create a life of purpose. He saw simplicity, when it is voluntarily chosen, as a vital ally in achieving our life purpose because it enables us to cut through the complexity and busyness of the world. Gregg asked us to consider: What is the unique and true gift that only you can bring to the world? Realizing your life-purpose—or using your true gift—will determine how you structure your life. For example, if your true gift is to adopt and raise a bunch of kids, then you may need to own a large house and car. If your true gift is creating art, then you may choose to forego the house and car and instead travel the world and develop your art. Simplicity is the razor's edge that cuts through the trivial and finds the essential. Simplicity is not about a life of poverty, but about a life of purpose. Here is a key passage from Gregg's writing that describes the essence of voluntary simplicity:

> Voluntary simplicity involves both inner and outer condition. It means singleness of purpose, sincerity and honesty within, as well as avoidance of exterior clutter, of many possessions irrelevant to the chief purpose of life. It means an ordering and guiding of our energy and our desires, a partial restraint in some directions in order to secure greater abundance of life in other directions. It involves a deliberate organization of life for a purpose. Of course, as different people have different purposes in life, what is relevant to the purpose of one person might not be relevant to the purpose of another . . . The degree of simplification is a matter for each individual to settle for himself.[4]

The more I thought about the phrase "voluntary simplicity," the more I appreciated its power. To live more *voluntarily* is to live more consciously, deliberately, and purposefully. We cannot be deliberate when we are distracted and unaware. We cannot be intentional when we are not paying attention. We cannot be purposeful when we are not being present. Therefore, to act in a voluntary manner is not only to pay attention to the actions we take in the outer world, but also to pay attention to the one who is acting—to our inner world.

To live more *simply* is to live more lightly, cleanly, aerodynamically—in the things that we consume, in the work we do, in our relationships with others, and in our connections with nature. We each know the unique distractions, clutter, and pretense that weigh upon our lives and make our passage through life needlessly difficult. In living more simply, we make our journey more easeful and rewarding.

Voluntary simplicity means living in such a way that we consciously bring our most authentic and alive self into direct connection with life. This is not a static condition, but an ever-changing balance. Simplicity in this sense is not simple. To live out of our deepest sense of purpose—integrating and balancing the inner and outer aspects of our lives—is an enormously challenging and continuously evolving process. The objective of the simple life is not to live dogmatically with less, but rather to live with balance so as to have a life of greater fulfillment and satisfaction.

There is no instruction manual or set of criteria that defines a life of conscious simplicity. Gregg was insistent that "simplicity is a relative matter depending on climate, customs, culture, and the character of the individual."[5] Henry Thoreau was equally clear that there is no easy formula defining the worldly expression of a simpler life. "I would not have anyone adopt my mode of living on my account . . . I would have each one be very careful to find out and pursue his own way."[6] Because simplicity has as much to do with our purpose in living as it does with our standard of living and because we each have a unique purpose in living, it follows that there is no single right and true way to live more ecologically and compassionately.

Drawn from my book *Voluntary Simplicity,* here are a few firsthand descriptions of this way of life, offered by people who are pioneers of living simply by choice:

"Voluntary simplicity has more to do with the state of mind than a person's physical surroundings and possessions."

"As my spiritual growth expanded and developed, voluntary simplicity was a natural outgrowth. I came to realize the cost of material accumulation was too high and offered fewer and fewer real rewards, psychological and spiritual."

"It seems to me that inner growth is the whole moving force behind voluntary simplicity."

"We are intensely family oriented—we measure happiness by the degree of growth, not by the amount of dollars earned."

"I feel this way of life has made my marriage stronger, as it puts more accent on personal relationships and inner growth."

"I consciously started to live simply when I started to become conscious."

"The main motivation for me is inner spiritual growth and to give my children an idea of the truly valuable and higher things in this world."

"I feel more voluntary about my pleasures and pains than the average American who has his needs dictated by Madison Avenue (my projection, of course). I feel sustained, excited, and constantly growing in my spiritual and intellectual pursuits."

"To me, voluntary simplicity means integration and awareness in my life."

What emerges from these descriptions is the sense that something intangible is essential to these people's lives. Perhaps it is living with a feeling of reverence for the Earth and all life, or cultivating a sense of gratitude rather than greed, or focusing on the quality and integrity of relationships of all kinds. At the heart of a life of conscious simplicity is some form of experiential spirituality. In contrast to the larger society where cynicism is rampant, this is a community of people who are tapping into, valuing, and trusting their felt experience of the sacred, although they describe that experience in many different ways.

Voluntary Simplicity and Soulful Living

WRITING in 1845, Henry Thoreau set the soulful tone for the simple life in *Walden,* in which he wrote these famous lines:

> I went to the woods because I wished to live deliberately, to confront all of the essential facts of life, and see if I could learn what it had to teach, and not, when I came to die, to discover that I had not lived. . . . I wanted to live deep and suck out all the marrow of life . . .[7]

The Hindu poet Tagore wrote, "I have spent my days stringing and unstringing my instrument while the song I came to sing remains unsung." Those choosing a life of simplicity are not leaving the song of their soul unsung. Instead, they are living "deep," diving into life with engagement and enthusiasm. And, in living that way, they are no doubt experiencing what Thoreau discovered—that "it is life near the bone where it is sweetest." To live simply is to approach life and each moment as inherently worthy of our attention and respect, consciously attending to the small details of life. In attending to these details, we nurture the soul. Thomas Moore explains in *Care of the Soul*:

> Care of the soul requires craft, skill, attention, and art. To live with a high degree of artfulness means to attend to the small things that keep the soul engaged . . . to the soul, the most minute details and the most ordinary activities, carried out with mindfulness and art, have an effect far beyond their apparent insignificance.[8]

For many, the American dream has become the soul's nightmare. Often, the price of affluence is inner alienation and emptiness. Not surprisingly, polls show that a growing number of Americans are seeking lives of greater simplicity as a way to rediscover the life of the soul.

Although the mass media may focus on the external trappings of a simple life, if we look below the surface, we find a powerful new form of personal spirituality motivating the vast majority of these lifeway innovators. For many, their spirituality is an individualized form of faith that minimizes rules and absolutes, and bears little resemblance to the pure form of any of the world's religions.[9] Their experience with the soulful dimensions of life and relationships is so rich and meaningful that a consumerist lifestyle appears pale by comparison.

I have had a quarter century of experience writing about,

speaking about, and living a life of voluntary simplicity. Based on that, here are other priorities (beyond material frugality) that I have found that characterize this way of living:

- **Relationships**—Those choosing the simple life tend to place a high priority on the quality and integrity of their relationships with every aspect of life—with themselves, other people, other creatures, the Earth, and the universe.
- **True gifts**—This way of living supports discovering and expressing the true gifts that are unique to each of us, as opposed to waiting until we die to discover that we have not authentically lived out our true potentials.
- **Balance**—The simple life is not narrowly focused on living with less; instead, it is a continuously changing process of consciously balancing the inner and outer aspects of our lives.
- **Meditation**—Living simply enables us to approach life as a meditation. By consciously organizing our lives to minimize distractions and needless busyness, we can pay attention to life's small details and deepen our soulful relationship with life.

All of the world's spiritual traditions have advocated an inner-directed way of life that does not place undue emphasis on material things. The Bible speaks frequently about the need to find a balance between the material and the spiritual sides of life, such as in this passage: "Give me neither poverty nor wealth" (Proverbs 30:8). From China and the Taoist tradition, Lao-tzu said that "he who knows he has enough is rich." In Buddhism, there is a conscious emphasis on discovering a middle way through life that seeks balance and material sufficiency.

The soulful value of the simple life has been recognized for thousands of years. What is new is that world circumstances are changing so that this way of life now has unprecedented relevance for our times.

THE SPRINGTIME OF SIMPLICITY

IN the 1960s, voluntary simplicity was a lifeway adopted by a handful of social mavericks; today, a little more more than thirty years later, it is a mainstream wave of cultural invention involving millions of people. Gerald Celente, president of the Trends Research Institute, reported in 1997 on how the voluntary simplicity trend is growing throughout the industrialized world: "Never before in the Institute's seventeen years of tracking has a societal trend grown so quickly, spread so broadly and been embraced so eagerly."[10] In the United States, a conservative estimate is that in the late 1990s, 10 percent of the adult population—or more than 20 million people—are opting out of the rat race of consumerism and into soulful simplicity.[11]

The following surveys provide further evidence that a lifeway of soulful simplicity, with its new pattern of values, is emerging as a significant trend in the world.

Yearning for Balance—A 1995 survey of Americans commissioned by the Merck Family Fund found that respondents' deepest aspirations are nonmaterial.[12] For example, when asked what would make them much more satisfied with their lives, 66 percent said, "if I were able to spend more time with my family and friends," and only 19 percent said, "if I had a bigger house or apartment." Twenty-eight percent of the survey respondents said that in the last five years, they had voluntarily made changes in their lives that resulted in making less money, such as reducing work hours, changing to a lower-paying job, or even quitting work. The most frequent reasons given for voluntarily downshifting were:

- Wanting a more balanced life (68 percent)
- Wanting more time (66 percent)
- Wanting a less stressful life (63 percent)

Had it been worth it? Eighty-seven percent of the downshifters described themselves as happy with the change. In summing up the survey's findings, the report states, "People express a strong desire for a greater sense of balance in their lives—not to repudiate material gain, but to bring it more into proportion with the nonmaterial rewards of life."

The Rise of Integral Culture—A random national survey conducted by Paul Ray in 1995 found that about 10 percent of the U.S. population (roughly 20 million adults) are choosing to live in a way that integrates a strong interest in their inner or spiritual life with an equally strong concern for living more in harmony with nature.[13] Ray calls these people "cultural creatives." As a group, they live more simply, work for ecological sustainability, honor nature as sacred, affirm the need to rebuild communities, and are willing to pay the costs for cleaning up the environment. As individuals, they are largely unaware of one another and feel relatively isolated.

World Values Survey—This massive survey was conducted in 1990–1991 in forty-three nations representing nearly 70 percent of the world's population and covering the full range of economic and political variation.[14] Ronald Inglehart, global coordinator of the survey, concluded that over the last twenty-five years, a major shift in values has been occurring in a cluster of a dozen or so nations, primarily in the United States, Canada, and Northern Europe. He calls this change the "postmodern shift."[15] In these societies, emphasis is shifting from economic achievement to postmaterialist values that emphasize individual self-expression, subjective well-being, and quality of life. At the same time, people in these nations are placing less emphasis on organized religion, and more on discovering their inner sense of meaning and purpose in life.[16]

Health of the Planet Survey—In 1993, the Gallup organization conducted in twenty-four nations a landmark global survey of attitudes toward the environment.[17] In writing about the survey, its director Dr. Riley E. Dunlap concluded that there is "virtually worldwide citizen awareness that our planet is indeed in poor health, and great concern for its future well-being." The survey found that residents of poorer and wealthier nations express nearly equal concern about the health of the planet. Majorities in most of the nations surveyed gave environmental protection a higher priority than economic growth, and said that they were willing to pay higher prices for that protection. There was little evidence of the poor blaming the rich for environmental problems, or vice versa. Instead, there seems to be a mature and widespread acceptance of mutual responsibility. When asked who is "more responsible for today's environmental problems in the world," the most frequent response was that industrialized and developing countries are "both equally responsible."

***World Environmental Law Survey*—**The largest environmental survey ever conducted was done in the spring of 1998 for the International Environmental Monitor. Involving more than thirty-five thousand respondents in thirty countries, the survey found that "majorities of people in the world's most populous countries want sharper teeth put into laws to protect the environment."[18] Majorities in twenty-eight of the thirty countries surveyed (ranging from 91 percent in Greece to 54 percent in India) said that environmental laws as currently applied in their country "don't go far enough." The survey report concludes, "Overall, these findings will serve as a wake-up call to national governments and private corporations to get moving on environmental issues or get bitten by their citizens and consumers who will not stand for inaction on what they see as key survival issues."

Could a shift to postmaterialist values occur rapidly if this reservoir of sympathy and support were encouraged? Could these social entrepreneurs be planting seeds of innovation for an evolutionary bounce several decades hence?

Although these global surveys show promising evidence of a shift from consumerism toward sustainability, it is not clear whether this shift will influence the newly modernizing economies of Africa and Asia. For example, in a Gallup survey conducted in China in October 1994, people were asked which attitudes toward life came closest to describing their own. Sixty-eight percent said that to "work hard and get rich" came closest to describing their approach to life, while only 10 percent selected "don't think about money or fame, just live a life that suits your own taste."[19] Clearly, consumerist attitudes are flourishing in Asia and are likely to come into conflict with the need to develop more ecological ways of living. Indeed, the trends toward sustainability in a number of postmodern nations could be overwhelmed by the impact of rapid industrialization in just two nations, China and India, with their combined population of roughly 2 billion people.

Implications for the Future

IF a new way of life does emerge that values simplicity and satisfaction over consumerism, the implications will be enormous. I believe they will include sustainable economic development, greater economic justice, new forms of community, greater participation in the political system, the development of human potentials, and the advancement of our civilizational purpose.

Sustainable Economic Development. Consumer purchases account for nearly two-thirds of the economic activity in the United States. If a significant percent of Americans were to change their consumption levels and patterns, the effects would be dramatic. Over the years, I have noticed that people choosing a simple life tend to make these kinds of changes in their consumption:

- They tend to buy products that are durable, easy to repair, nonpolluting in their manufacture and use, energy-efficient, not tested on animals, functional, and aesthetic.

In addition, they are more inclined to make their own furniture, clothing, and other products as a form of self-expression.

- Regarding transportation, people choosing a life of simplicity tend to use public transit, car-pooling, bicycles, and smaller and more fuel-efficient cars; they may walk rather than ride; they often live closer to work; and they tend to make more extensive use of electronic communication and telecommuting as a substitute for physical travel.

- They often pursue livelihoods that contribute to others and enable them to use their creative capacities in ways that are fulfilling.

- They tend to shift their diets from highly processed food, meat, and sugar toward foods that are more natural, healthful, simple, locally grown, and appropriate for sustaining the inhabitants of a small planet.

- They recycle metal, glass, plastic, and paper and cut back on their use of things that waste nonrenewable resources.

- They reduce undue clutter and complexity in their lives by giving away or selling things that they seldom use, such as clothing, books, furniture, and tools.

- They tend to buy less clothing, jewelry, and cosmetics; they tend to focus on what is functional, durable, and aesthetic rather than on passing fads, fashions, and seasonal styles.

- They usually observe holidays in a less commercialized manner.

Bit by bit, these and other small changes by individuals and families could coalesce into a tremendous wave of economic change in support of a sustainable future. Professor Stuart Hart, writing in the *Harvard Business Review* about strategies for a sustainable world, says that "over the next decade or so, sustainable development will constitute one of the biggest opportunities in the history of commerce."[20]

How would a sustainable economy differ from a consumer economy? For one thing, it would be much more differentiated: some sectors would contract (especially those that waste energy and are oriented toward conspicuous consumption), while other sectors would expand (such as information processing, interactive communications, intensive agriculture, retrofitting homes for energy efficiency, and education for lifelong learning). To minimize the costs of transportation and distribution, markets would be more decentralized than they are today. People would buy more goods and services from local producers; in turn, there would be a rebirth of entrepreneurial activity at the local level. Small businesses well adapted to local conditions and needs would flourish. New types of markets and marketplaces would proliferate, such as flea markets, community markets, and extensive bartering networks (whose efficiency will be greatly enhanced by new generations of computers that match goods and services with potential consumers or traders). The economy would also be more democratized as workers take a larger role in decision making. All types of products—such as cars, refrigerators, and carpeting—would be designed to be easily disassembled and then recycled into new products, minimizing waste. Less money would be spent on material goods and more on entertainment, education, and communication.

One criticism of the simple life is that it would undermine economic growth and produce high unemployment. This criticism is based on the erroneous assumption that high-consumption lifestyles are necessary to maintain a vigorous economy and full employment. However, in modern consumer societies such as the United States, there are an enormous number of unmet needs—for example, restoring the natural environment, retrofitting our homes for sustainable living, rebuilding our decaying cities, caring for the elderly, and educating the young. For the foreseeable future, there will be no shortage of real work and meaningful employment if we are committed to meeting the real needs of people.

Likewise, in developing nations, there is enormous economic opportunity if approached from the mind-set of sustainability. Sixty percent of the world's population lives on the equivalent of $3 or less a day, mostly in the developing world, in urban shantytowns without adequate shelter, clean water, sanitation, schools, health care, fire and police protection, access to communications technology, dependable energy, paved roads, public transportation, or space to grow food. These enormous needs represent equally great economic opportunities for meaningful work.

Economic justice. The Universal Declaration of Human Rights affirmed by the United Nations in 1948 states in part that "everyone has the right to a standard of living adequate for the health and well-being of himself and of his family, including food, clothing, housing and medical care and necessary social services." A significant part of humanity has no way to exercise that right, and I see little possibility of that changing under the trickle-down economic system we have today.

Given the new perceptual paradigm that is emerging—

whose core expression is a shift in experience from existential separation in a dead universe to empathic connection in a living universe—it is not surprising that those who choose a simpler life tend to feel connected with and a compassionate concern for the world's poor. This sense of kinship with people around the world fosters a concern for social justice and greater fairness in the use of the world's resources. Because economic inequality is increasing rapidly in the world, a conscious cultural shift toward more sustainable levels and patterns of consumption seems essential if there is to be greater equity in how people live. Indeed, I see a lifeway of choiceful simplicity and graceful moderation as the only realistic foundation for achieving a meaningful degree of economic fairness and thereby building a foundation for pulling together as a human family.

We need to learn to use resources more fairly if we are to live peacefully. Armies and military weapons are enormously expensive and represent a huge drain on resources that could otherwise be used for sustainable development. If we are able to narrow the gap between the rich and the poor of the world, the prospect of conflict over scarce resources will diminish. This, in turn, could free up people and resources for building a future that benefits us all.

New forms of community. Community provides the foundation for a civilization of simplicity. To encourage self-reliance at the most local scale feasible, community design would likely involve a nested set of living arrangements. For example, a family would live in an "eco-home" (designed for considerations such as energy efficiency, telecommuting, and gardening), nested within an "eco-neighborhood," within an "eco-village," within an "eco-city," within an "eco-region," and so on. Each eco-village could contain a telecommuting center, child-care home, community garden, and recycling area. Urban land that was formerly used for lawns and flower gardens could be used for supplemental food sources such as vegetable gardens, and fruit and nut trees. These micro-communities or neighborhood-size villages could have the flavor and cohesiveness of a small town combined with the urban flavor of a larger city. Each eco-village might specialize in a particular kind of work—such as crafts, health care, child care, gardening, or education—providing fulfilling work for many of its inhabitants. People could earn time-share hours that could be bartered for the products or services of neighbors—such as gardening, food, music lessons, carpentry, or plumbing. People could balance their work between serving their local community and serving the world.

Because the populations of eco-villages (five hundred or so people) would approximate the scale of a tribe, many people could feel quite comfortable in this design for living. With an architecture sensitive to the psychology of these modern tribes, a new sense of community could begin to replace the alienation of today's massive cities. To support these innovations in housing and community, there could be accompanying changes in zoning laws, building codes, financing methods, and ownership arrangements. Overall, these smaller-scale, human-size living and working environments could foster a rebirth of community; we could again have face-to-face contact in the process of daily living in local neighborhoods.

Greater participation in politics. Many of those choosing a simpler way of life have pulled back from traditional politics, unable to identify with either conservatives (who tend to trust in the workings of business and the marketplace) or liberals (who tend to trust in the workings of government and bureaucracy). They are turning instead to their own resources as well as to their friends and local community. The politics of simplicity are neither left nor right, but represent a new combination of self-reliance, community spirit, and cooperation.

We can use the analogy of humanity's adolescence to get a better sense of how politics may change in the future as we mature into our young adulthood. It seems to me that humans have been acting like political adolescents; on the whole, we have been waiting for "Mom and Dad" (our big institutions of business and government) to take care of things for us and we blame them when they don't. As we move into our early adulthood, however, we are beginning to face our challenges head on, recognizing that we are in charge, and that no one is going to save us. To create a sustainable future for ourselves on this planet, particularly given the speed, cooperation, and creativity that our situation demands—will require the voluntary actions of millions, even billions, of free individuals acting responsibly and in concert with one another. Never before in human history have so many people been called upon to make such sweeping changes voluntarily and in so little time. The new politics are grounded in the unflinching recognition that we are being challenged to grow up and take charge of our lives, both locally and globally.

Our indispensable ally in this process is the communications revolution. When the politics of sustainability are combined with the power of television and the internet, the combination could be transformative. As we shall explore in the next chapter, the communications revolution will support a dramatic increase in the public's ability to hold corporations and governments accountable for their actions. Internet campaigns will flourish that blow the whistle on government and corporate abuses and encourage people to boycott the products of firms and nations whose policies are unethical environmentally, economically, or socially.

Finally, a new era of volunteerism could blossom. For instance, young people could be encouraged to contribute a year or more of local or national service, perhaps restoring the environment, working with youth, or building community centers.

The development of human potentials. A life that is outwardly simple and inwardly rich naturally celebrates the development of our many potentials. As the simple life makes time available, areas for learning and growth blossom. These include the physical (such as running, biking, and yoga); the emotional (such as learning the skills of emotional intelligence and interpersonal intimacy); the intellectual (such as developing skills in the arts and crafts as well as basic skills such as carpentry, plumbing, appliance repair, and gardening); and the spiritual (such as various forms of meditation and relaxation, and exploring the mind-body connection with biofeedback).

The advancement of our civilizational purpose. Choosing to live more simply does not mean turning away from progress; quite the opposite. A life-way of voluntary simplicity is a direct expression of our growth as a maturing civilization. After a lifetime of studying the rise and fall of more than twenty of the world's civilizations, the highly esteemed historian, Arnold Toynbee, concluded that the conquest of land or people was not the true measure of a civilization's growth. The true measure, he said, was expressed in a civilization's ability to transfer an increasing proportion of energy and attention from the material to the nonmaterial side of life to develop its culture (such as music, art, drama, and literature), sense of community, and strength of democracy. Toynbee called this the "Law of Progressive Simplification."[21] He said that authentic growth consists of a "progressive and cumulative increase both in outward mastery of the environment and in inward self-determination or self-articulation on the part of the individual or society."[22] I believe that Toynbee is correct, and that our outward mastery will be evident by living ever more lightly upon the Earth, and our inward mastery will be evident by living ever more lightly with gratitude and joy in our hearts.

Choosing a way of life that is simpler, more satisfying, and more sustainable could help us transform an evolutionary crash into a bounce. Obviously, the simple life offers no magical solutions. It will take millions and even billions of people tending to the small details of their lives to craft a more soulful and satisfying existence for themselves and for us all. It is, nonetheless, empowering to know that each of us can make a meaningful difference by taking responsibility for changes in our own lives. Most of us have seen the limits of bureaucracy and understand that, if creative action is required, it will likely come through the conscious actions of countless individuals working in cooperation with one another. A lifeway of conscious simplicity is made to order for self-organizing action at the local scale. Small changes that seem insignificant in isolation can have an enormous impact when undertaken together by millions.

Seeds growing in the garden of simplicity for the past generation are now blossoming into the springtime of their planetary relevance and could provide a crucial ingredient for an evolutionary bounce.

Chapter Five

Communicating Our Way into a Promising Future

The communications industry is the only instrument that has the capacity to educate on a scale that is needed and in the time available.

—*Lester Brown, President of Worldwatch Institute*

Awakening as a Planetary Species

OUR ability to communicate has enabled humans to progress from nomadic bands of gatherers and hunters to the edge of a planetary civilization. Thus it should come as no surprise that our ability to communicate will determine whether we are successful in achieving a promising future. Anything that dramatically enhances human communication will have an equally dramatic influence on our evolution. Because we are in the midst of an unprecedented revolution in the scope, depth, and richness of global communications, the impact of this revolution on our future will be equally unprecedented. The global communications revolution is such an extraordinary force for change in the world that I consider it one of the key factors that could transform an evolutionary catastrophe into an exhilarating leap forward for humanity.

Prior to the era of electronic communications, the world

was a vast place where oceans and continents insulated people from one another. Events in one part of the world might be utterly unknown elsewhere for months or years, if they were ever known at all. Over the last few decades, the change in our ability to communicate has been extraordinary. To illustrate just how much things have changed, I'll use my own experience as an example.

In the 1950s, I was in my early teens, living on a family farm in Idaho, several miles from a town of five hundred people. I felt disconnected from the larger world, as a couple of radio stations were my primary contact with life beyond our small town. Television was just arriving and consisted of three channels broadcasting snowy, black-and-white images during the afternoons and evenings. The weekly paper was a scant few pages covering the news of a small town, so I looked forward to the weekly edition of the *Saturday Evening Post* and the monthly *National Geographic*. To a great extent, the events occurring around the world were largely unknown, and unknowable to me.

One of the most exciting events during those years occurred one winter when I built my first shortwave radio from a kit. For several days, I soldered together resistors, capacitors, tubes, and transformers. Then I went outside and hooked an antenna wire from the roof of our house to a telephone pole some thirty yards away. On a cold and clear winter night, I turned on my shortwave receiver to listen in on the world. After a few minutes of turning the dial, a voice came in distinct and strong—from Australia! I was completely amazed to be listening to someone speaking from the other side of the world. Hearing that voice suddenly made the Earth feel much smaller and more approachable.

What was miraculous to me then is now commonplace

given the explosive growth in communications technologies. Whereas I was able to listen to a random voice by accident, people can now interact purposefully and inexpensively with countless other people around the planet. The internet is collapsing the world into an electronic village where we are all neighbors, while television is providing a common world-language through its visual images.

The spectacular growth in global communications offers humanity the possibility of communicating our way through this time of planetary challenge. All of the adversity trends that were discussed in chapter two—such as climate change, population growth, poverty, the extinction of species, and the depletion of fresh water—are, at their core, communications challenges. For example, coping with climate change is not primarily a technical issue concerning carbon dioxide in the atmosphere; more fundamentally, it is about our ability to communicate together as a global family and respond in concert in ways that will reduce emissions. Overpopulation is, in many ways, a more fundamental problem of illiteracy and lack of educational opportunity for women—again, a communications challenge. Responding to global poverty represents an enormous communications challenge as we try to reach a working understanding of how to live fairly in a communications-rich world where inequities have become glaringly obvious. Ensuring that there is enough food, water, and fuel for humanity's future is not simply a material concern, it is also a matter of planning ahead on a planetary scale, and this is a communications challenge. Which energy future we take—relying on decentralized sources of renewable energy such as solar and wind, or relying on centralized sources such as nuclear power—is also a communications issue. It seems very hopeful to me that we are fast acquiring the communication tools needed to talk through all of these challenges.

The two most powerful expressions of the communications revolution are television and the internet. Let's look at each of these media briefly.

Television: The Pathway to Collective Visualization

TELEVISION has been called "the boob tube," "a vast wasteland," and "a golden goose that lays scrambled eggs," yet it is an incredibly powerful medium. It is powerful because it makes use of visual communication, and we are a visually oriented species. (We don't say things like "seeing is believing" and "one picture is worth a thousand words" for no reason.) The visual imagery of television creates a common language that makes it humanity's primary source of shared information and understanding. Television is the window through which we see the world, and the mirror in which we see ourselves.

A key measure of the power of television in today's world can be seen when there is a military challenge to the authority of a national leader. In decades past, during a military coup, the army would seek to take over the train stations and airports, as these were the main conduits of power. No longer. In a showdown of power today, the focus of military attention as well as civilian attention are the television stations. In the recent era, when there has been the threat of civil war—whether in Russia, the Philippines, or Eastern Europe—the armies and civilians have gathered primarily around the television stations and towers, recognizing these are the real seats of power as they provide the all-important, visual connection with the collective consciousness of the society.

In the United States, 98 percent of all homes have a TV set, which the average person watches more than four hours per day. Taken together, this means that Americans watch approximately one billion person-hours of television daily. A majority of Americans get a majority of their news about the world from this medium. There are more homes in the United States with a TV set than with indoor toilets, stoves, or refrigerators. A similar emphasis on televised communication exists elsewhere around the world. In China, for example, just 2 percent of homes in major cities had hot running water in 1997, but 89 percent had televisions.[1] In India, South America, and other developing regions, many people who lack indoor plumbing, refrigerators, and other basics nonetheless have a TV set connecting them with the world. Despite enormous disparities in material wealth, people in developing nations are already part of the global communications culture.

The closeness and intimacy of television's window onto the world can give people a feeling of connection with the fate of the Earth. At the speed of light, television can unite the entire planet. Through the eyes of television, we can see a starving child in Africa, the destruction of rain forests in Brazil, urban decay in New York City, and the effects of acid rain in Germany. Television makes every viewer an active witness—a knowing and feeling participant in what is being shown. Professor of communications George Gerbner recognized the power of television as the primary storytelling machine of civilizations. He said that to control a nation you don't have to control its laws or its military, but rather control who tells its stories. Television tells most of the people most of their stories most of the time. And the people who now control television are advertisers despite the fact that it is the public who "own the airwaves."

Whether used for good or ill, the influence of television often overwhelms the power of religious institutions, its capacity to shape public opinion surpasses that of many political institutions, and its ability to program the minds of children is stronger than that of our schools. Television is our social witness, our vehicle for "knowing that we know" as communities, nations, and as a human family. If issues and concerns do not appear regularly on television, then for all practical purposes they do not exist in mass social consciousness. Television is now an integral part of the social brain of our species.

The Internet: Pathway to the Global Brain

WHERE television has overcome language barriers with the power of visual communication, the internet has overcome distance barriers with the power of its planetary reach. The internet—with its billions of cross-cultural communications and connections flowing daily around the world—has transformed political and social boundaries into permeable membranes. Individuals are gaining instant access to one another, and are communicating in ways formerly reserved only for the very wealthy. A communications-rich future holds the promise of the radical empowerment of humanity at the grassroots level. Web observer Mark Pesce has written that "it is not an overstatement to frame the World Wide Web as an innovation as important as the printing press—it may be as important as the birth of language itself . . . in its ability to completely refigure the structure of civilization."[2]

The raw power of internet technologies is like nothing we have known before. According to Joseph Pelton, who has written extensively about the globalization of telecommunications, a single advanced satellite or fiber-optic cable currently has the capability of sending the entire *Encyclopedia Britannica* with all its illustrations every three seconds.[3] Future prospects are even more breathtaking: "In another quarter of a century these are likely to be . . . systems that could send the equivalent of the entire U.S. Library of Congress in less than 10 seconds."[4] Another observer of the internet is John Midwinter, who has written that the computing and communications industry "shows every sign of continuing its breathtaking pace for at least one or two decades more (e.g., doubling performance every one or two years), implying a revolution in capability every five to ten years."[5] Growth of the internet supports this claim. Worldwide, an estimated 40 million people used the internet in 1996. This number is expected to jump to 500 million by 2001.[6] By 2010, there will be an estimated one billion people continuously connected to the World Wide Web.[7]

When this planetary scope of human connection is combined with the functional intelligence of computers, a new level of human awareness and communication—a "global brain"—could potentially emerge. What this means is that we could soon achieve a quantum increase in the functional intelligence of our species.[8] Billions of messages are swirling around the globe each day, weaving the human family into an ever-tighter web of communication and consciousness. With the connection of billions of people into a single system, a rudimentary global brain is emerging.[9] When a billion people or more are connected into an integrated communication network twenty-four hours a day, seven days a week, a new level of collective consciousness with unexpected potentials will awaken in the world.

THE QUICKENING PACE OF GLOBAL AWAKENING

HOW soon might the emerging global brain reach a critical threshold and turn on with a rudimentary collective consciousness? Research by the Institute for Information Studies states that the physical infrastructure of the global communications network that will serve as the conduit for widespread economic, social, cultural, and political exchange will "start to come into place around the second decade of the twenty-first century."[10] This is same time frame in which the adversity trends discussed in chapter two are expected to converge into a whole-systems crisis for the people of the Earth.

Because fiber-optic cable will likely be the medium of choice for high-density routes in developed countries, the rate of its use is one meaningful indicator of the rate at which the global brain is being wired. Projections by Bell Northern Research for the growth and deployment of optic fiber in developed countries indicate that by the mid-2020s, the basic communications infrastructure will be in place to support a major leap forward in our ability to communicate.[11]

Expanding the reach of the internet, hundreds of satellites will be launched over the next few years to form the infrastructure for a wireless communications system. Writing in *Scientific American,* Joseph Pelton predicts that by 2003, there will likely be one thousand commercial communications satellites in service, up from about two hundred twenty in 1998.[12] This new generation of satellites, placed in low Earth orbit, will revolutionize global communications. For better or worse, many people will soon be continuously connected with the world no matter where they are.

Combined with satellites, cellular phones will enable developing countries to bypass the need to build a vast network of telephone lines strung along poles. There is an enormous advantage to this—it enables developing countries to literally leapfrog into the future, by avoiding investment in traditional forms of equipment and communication networks.[13] The internet could also have a very positive impact on developing countries.[14] For some people, it offers the opportunity of global telecommuting. There are now software programmers in India, for example, who telecommute daily to work in the Silicon Valley. For others, the internet offers tele-medicine—low-cost, online, medical assistance—even in remote areas of the world. The internet can also help isolated groups find markets for goods and services, and empower local activists by linking them with supporters across the globe.

Given all these communication trends, it seems likely that *within ten to twenty years, we will have in place the communications infrastructure that could support a quantum increase in the collective communication—and the collective consciousness—of our species.*

I do not assume that electronic communication can or should carry the entire burden of human communication. It is vital that we combine the power of global communication with study circles and other forms of grassroots dialogue. With the combination of the two, a local-to-global conversation could emerge to shape the outlines of a sustainable future. Individuals could engage in face-to-face conversations locally—in homes, schools, churches, civic organizations, and workplaces—and then connect with local groups around the world via the internet. If we could generate this kind of worldwide dialogue, it seems plausible that the human family could mobilize itself to begin building a future that we scarcely could have imagined a decade or two earlier.

The Abuse of Power—Perceptual Totalitarianism

LIKE all powerful technologies, the tools of global communication present humanity with a double-edged sword. On the one hand, these are the tools that we need to build a sustainable future. On the other hand, these are also the tools that are now being used to promote mindless consumerism around the planet—which may be the primary threat to a sustainable future.

In the course of a year, the average American will view approximately twenty-five thousand commercials. A commercial represents far more than a pitch for a particular product; it is also an advertisement for the attitudes, values, and lifestyle that surround the consumption of that product. The unrelenting consumerist bias of television distorts our view of reality and social priorities, leaving us entertainment rich and knowledge poor. Television may be our window onto the world, but the view it now provides is cramped and narrow. Television may be the mirror in which we see ourselves as a society, but the reflection it gives is distorted and unbalanced.

The mirror that mainstream television holds up to the world as the realistic norm of consumption is far removed from the lives of a majority of people on the Earth. The use of television to promote primarily materialistic values has become a severe, although largely unacknowledged, mental health and public health problem for the United States and the world. We are in a double bind: while the mass media that dominate our consciousness tell us to consume more, our ecological concerns tell us to consume less. Carl Jung defined schizophrenia as a condition where "the dream becomes the reality." The American dream of a consumerist way of life has become a dangerous illusion that no longer fits the reality of the world and our human potentials. By allowing a commercialized view of the world to become our primary way of defining reality, we are becoming a schizophrenic society. As commercial television spreads throughout the world, with it grows a deep conflict within our collective psyche.

A major survey of American college freshmen over a period of thirty years gives us striking evidence of the powerful impact of television on values. This study, by the American Council on Education and UCLA, found that there has been a dramatic shift in the values of college freshmen since the 1960s.[15] In 1966, "developing a meaningful philosophy of life" was the top value, being endorsed as a "very important" or "essential" goal by more than 80 percent of the entering freshmen. "Being well-off financially" lagged far behind, ranking fifth on the list, with less than 45 percent of freshmen endorsing it as a very important or essential goal in life. Since then, these two very different values have essentially traded rankings, as Figure 5 shows. In 1996, being well-off financially was the top value (74 percent of freshmen iden-

FIGURE 5

Contrasting Value Trends of U.S. College Freshmen

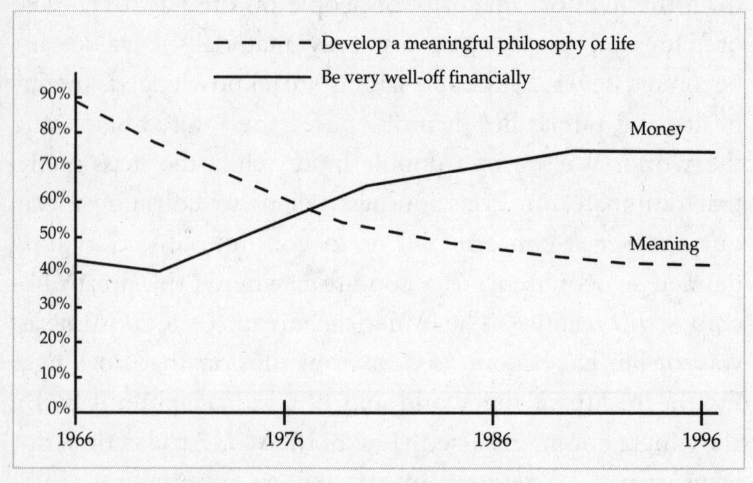

Source: American Council on Education and UCLA

tified it as very important or essential) and developing a meaningful philosophy of life fell to sixth place (only 42 percent named it as very important or essential).

According to the researchers who conducted this study, a major reason for this profound shift is the impact of television viewing on values. They found that "the more television watched, the stronger the endorsement of the goal of being very well off financially, and the weaker the endorsement of the goal of developing a meaningful philosophy of life."[16] Freshmen entering college in the late 1960s had been exposed to much less television while they were young than had freshmen entering college in the late 1980s. Given how ubiquitous the mass media have become since the 1960s, all generations henceforth will be swimming in increasingly sophisticated forms of television and thoroughly influenced by it.

I believe there may be no more dangerous challenge to our future than the cultural hypnosis that is generated daily by commercial television, which trivializes the human experience and distracts humanity from our larger potentials. *By programming television for commercial success, we are programming the mind-set of entire civilizations—indeed, the species-civilization—for evolutionary stagnation and ecological failure.* If television is our social brain, then American television currently provides the highest level of intelligence that beer and car commercials can buy. This dumbing down of our collective intelligence is happening at the very time that we face unprecedented upheaval and change. Our evolutionary maturity is being tested. Our future as a species may well depend on whether we take self-organizing action to ensure that our tools of collective communication support the social conversation required to build a sustainable future.

More than twenty years ago, author Gene Youngblood warned that the mass media could hold back human evolution simply by controlling the perception of alternatives. He said that in order to perpetuate the status quo, it is not necessary that the mass media create a genuine desire for a consumerist way of life; all they have to do is prevent genuine desire for any other way of life from being publicly expressed and collectively affirmed:

> Desire is learned. Desire is cultivated. It's a habit formed through continuous repetition . . . But we cannot cultivate that which isn't available. We don't order a dish that isn't on the menu. We don't vote for a candidate who isn't on the

> ballot . . . What could be a more radical example of totalitarianism than the power of the mass media to synthesize the only politically relevant reality, specifying for most people most of the time what's real and what's not, what's important and what's not . . . ? This, I submit, is the very essence of totalitarianism: the control of desire through the control of perception.[17]

By excluding visions of more sustainable ways of living and consuming, the mass media perpetuate the status quo while simultaneously crippling society's capacity for envisioning more promising alternatives. The experience of the Soviet Union provides a powerful example of what happens when people cannot consider alternatives via the public media. Historically, even though a majority of Soviet citizens were opposed to party-boss Communism, this opposition did not lead to radical reform of the social system because citizens were denied the opportunity to publicly consider alternatives to Communism. They knew what they were against, but they did not know what they were for. As one observer of the Soviet experience explains, "there was no 'model' of such reformation in the social consciousness."[18] Just as the Soviets have suffered in their transition because they lacked publicly considered alternatives to communism, Americans are suffering—and so is the world—because of an absence of publicly considered alternatives to consumerism.

As citizens of maturing societies, it is time to pay attention to how we pay attention collectively. It is time to break through the perceptual totalitarianism of the consumerist mind-set that undermines our future, and give ourselves alternative images of a sustainable future as vivid and compelling as the consumer messages that dominate our collective consciousness day after day. Only with sustained communication about alternative visions of the future can citizens consciously visualize, discuss, reflect upon, select, and build the future we want.

A closely related concern is the degree to which communications technologies become available in developing countries. If communications technologies continue to be another factor that separates the "haves" from the "have-nots," humanity will become increasingly divided into the "communications rich" and "communications poor," and this could foster further division and breakdown. Paradoxically, the fast developing, global communications network is making these economic divisions transparent to the rest of the world. For example, in the early 1990s, villagers in India could receive four channels of television, largely government controlled. By the late 1990s, the number of channels that could be received jumped to fifty-two and included a flood of U.S. and European programming. With that increase in channels has come a continuous stream of TV programs concerned with the lifestyle problems of middle-class people in developed countries. Because a majority of people in developing countries are seeing ways of living that are completely out of their reach for the foreseeable future, it would not be surprising if they were to feel resentment that comparable opportunities are not available to them and their children.

If only a minuscule fraction of people living in abject poverty in developing nations were to choose violence as the weapon of last resort, the impacts could be enormous in our interdependent world. To be heard, some among the disenfranchised may forgo physical violence only to embrace other forms of destruction. Warfare could shift to an electronic battleground, where terrorists could scramble or damage data critical to vital computer systems in government and business. Conflicts could range from minor skirmishes to all-out assaults on the electronic integrity of corporations, nations, or the entire planet. A recent estimate by information warfare specialists at the Pentagon reveals how vulnerable developed nations are to disruption. They estimated that a properly prepared and well-coordinated attack by fewer than thirty computer virtuosos, strategically located around the world, with a budget of less than $10 million, could bring the United States to its knees, shutting down everything from electric power grids to air traffic control centers.[19]

Power in a Transparent World

IN describing democracy in America in the early 1800s, Alexis de Tocqueville said that newspapers were a vital force, as they could put a single idea into ten thousand minds all on the same day. Less than two hundred years later, we have tools of communication that can put a single idea into several billion minds all at the same instant. The question is: What mixture of ideas and stories are we going to put into our collective consciousness?

How we use our tools of mass communication is not just another issue, it is the basis for understanding and responding to all issues. Through the internet, for example, we can quickly harness the expertise of the world to re-create our lives for sustainability. We can learn about solar technology for heating, photovoltaic technology for generating electricity, intensive urban gardening for supplemental sources of food, and so much more. New forms of business will emerge, providing services and products ranging from renewable energy systems to organic agriculture, from telemedicine to clothing and crafts from around the world. Nearly frictionless markets will facilitate barter networks, and will give people in developing countries much easier entry into global commerce.

The mass media and the internet are creating a transparent world where injustices are increasingly difficult to hide. This transparency is bringing a new level of accountability and ethicality into institutional conduct. There is no longer a rug under which corporations can sweep unjust labor practices, or governments can hide human rights abuses. Just as a rising tide lifts all boats, so too will a rising level of global communication lift all injustices into the healing light of public awareness.

What we see and hear through the new media may challenge the emotional intelligence and maturity of the species. As our social consciousness awakens, deep psychic wounds will emerge that have festered through history. We will begin to hear the voices that we have ignored and the pain that we have not acknowledged. Awakening may bring with it "the psychic sludge of history" in the form of racism, ethnic conflict, and religious discord.[20] It may seem unwise to bring the dark side of our past into the light of day, but, unless we do, this unresolved pain will forever pull at the underside of our consciousness and diminish our future potentials.

For the first time in human history, we are acquiring a way to listen to and talk with one another as members of one family. For the first time, all the cousins in the human clan can communicate with one another. In awakening to ourselves as a planetary species and seeing ourselves directly and whole for the first time, we will see that we have the potential for an evolutionary bounce. The question is whether we will have the collective maturity to seize this precious opportunity.

Chapter Six

Reconciliation and the Transformation of Human Relations

Love is mankind's most potent weapon
for personal and social transformation.
—*Martin Luther King, Jr.*

To him in whom love dwells,
the whole world is but one family.
—*The Buddha*

The Power of Love

IN the preceding chapters, we explored three trends that have the power to fuel an evolutionary bounce. One is the power of perception and the emergence of a perceptual paradigm that allows us to see the universe as alive rather than dead. Another is the power of choice and our ability to shift voluntarily toward a simpler way of life. A third is the power of communication and the opportunity to use the internet and the mass media to support a quantum increase in conversation about our common future. This chapter is about the fourth trend that has the capacity to transform a crash into a bounce—the power of love. By this I mean not romantic love, but a mature and soulful compassion that looks beneath surface differences and sees our common connection with the community of life.

Compassionate love is a transformative power that we cannot quantify or measure, yet it brings incomparable strength and resilience into human relationships. "Love," said Teilhard de Chardin, "is the fundamental impulse of Life . . . the one natural medium in which the rising course of evolution can proceed."[1] Without love, he said, "there is truly nothing ahead of us except the forbidding prospect of standardisation and enslavement—the doom of ants and termites."

A compassionate love can provide a vital "social glue" to hold us together as we face the challenges ahead. If we pull apart, an evolutionary crash seems assured. If we come together authentically, however, we have the real potential to achieve an evolutionary bounce. And to pull together, we need to reconcile the many differences that now divide us. We need to discover harmony where there is now discord. We need to cultivate the respect and regard for others that ultimately come from a foundation of love.[2]

Love is the deepest connecting force in the universe, and thus a vital ingredient in our evolutionary journey toward wholeness. The unfolding of love is not different from the unfolding of awareness. Jack Kornfield, esteemed meditation teacher, put it this way: "I will tell you a secret, what is really important . . . true love is really the same as awareness. They are identical."[3] If we can learn the lesson that love will further our evolution and that the greater our love the greater our awareness, then we are aligned for success in our journey home. With love—or a maturing awareness—as a foundation, the hallmark of the emerging era could be the healing of our many fragmented relationships. If that were to occur, it truly is possible to imagine a future that works for everyone. With reconciliation, there is little doubt that an evolutionary bounce could happen.

A compassionate or loving consciousness has ancient roots, but it is taking on a new importance as our world becomes integrated ecologically, economically, and culturally. Because we now share one another's fate, it is increasingly clear that promoting the well-being of others directly promotes our own. We have reached the point where the Golden Rule is becoming essential to humanity's survival. This ancient ethic, which is found in all of the world's spiritual traditions, advises that the way to know how to treat others is to treat them as you would want to be treated. Here are some of the ways the Golden Rule has been expressed:

> "As you wish that men would do to you, do so to them."
>
> —Christianity (Luke 6:31)

> "No one of you is a believer until he desires for his brother that which he desires for himself."
>
> —Islam (Sunan)

> "Hurt not others in ways that you yourself would find hurtful."
>
> —Buddhism (Udanavarga)

> "Do naught unto others that which would cause you pain if done to you."
>
> —Hinduism (Mahabharata 5:1517)

> "Do not unto others what you would not have them do unto you."
>
> —Confucianism (Analects 15:23)

As diverse and divisive as we are, the human family recognizes this common ethic of compassion at the core of life. This indicates to me that there is a basis for reconciliation within humanity.

Love and compassion not only have ancient roots; history also attests to their impact and enduring power. Compassionate teachers through the ages such as Jesus, Buddha, Mohammed, and Lao-tzu have all lacked wealth, armies, and political position. Yet as the late Harvard professor Pitirim Sorokin explains in his classic book, *The Ways and Power of Love,* they were warriors of the heart, and have reoriented the thinking and behavior of billions of people, transformed cultures, and changed the course of history: "None of the greatest conquerors and revolutionary leaders can even remotely compete with these apostles of love in the magnitude and durability of the change brought about by their activities."[4] In contrast, most empires built rapidly through war and violence—such as those of Alexander the Great, Caesar, Genghis Khan, Napoleon, and Hitler—have crumbled within years or decades after their establishment.

The ruler Ashoka, who lived in India three hundred years before Jesus was born, is an example of the power of love in human affairs.[5] Prince Ashoka was born into a great dynasty of warriors and inherited an empire that extended from central India to central Asia. Nine years into his reign, he launched a massive campaign to win the rest of the Indian subcontinent. Finally, after a fierce battle in which more than a hundred thousand soldiers were slain, the land was conquered. Ashoka walked the battlefield that day, looking at the dead and maimed bodies, and felt profound sorrow and regret for the slaughter and for the deportation of people he had conquered. He immediately ceased his military campaign, converted to Buddhism, and devoted the rest of his life to serving the happiness and welfare of all.

Ashoka's thirty-seven years of benevolent rule left a legacy of concern not only for human beings but for animals and plants as well. His decrees creating sanctuaries for wild animals and protecting certain species of trees may be the earliest example of environmental action by a government.[6] Ashoka's works of charity included planting shade trees and orchards along roads, building rest houses for travelers and watering sheds for animals, and giving money to the poor, aged, and helpless. His political administration was marked by the end of war and an emphasis on peace. All his political officers were encouraged to extend goodwill, sympathy, and love to their own people as well as to others. One of their main duties was to be peacemakers, building mutual goodwill among races, sects, and parties. His cultural activities promoted education and the arts of the stage, including the construction of amphitheaters. Sorokin sums up Ashoka's legacy as "a striking example of a peaceful, love-motivated, social, mental, moral and aesthetic reconstruction of an empire."[7]

Ashoka's compassionate rule established the largest kingdom in India until the arrival of the British more than two thousand years later. The lion pillar—a statue that is the symbol of Ashoka—survives to this day as the official emblem of the Republic of India, and is found on nearly every Indian coin and currency note. "Amidst the tens of thousands of names of monarchs that crowd the columns of history," wrote historian H. G. Wells, "the name of Ashoka shines, and shines almost alone, a star."[8]

Based on examples such as these, Sorokin concluded that love-inspired reconstructions of society carried out in peace are far more successful and yield much more lasting results than reconstructions inspired by hate and carried out with violence. Again and again, he found that "hate produces hate, physical force and war beget counterforce and counterwar, and that rarely, if ever, do these factors lead to peace and social well-being."[9]

Reconciling Our Many Divisions

IN the industrial era, extended relationships and traditional communities have been torn apart. We have become a species marked by rootless, mobile societies, in which temporary relationships and friendships predominate. Friends come and go. Marriages break up. People move away from their parents and siblings, sometimes geographically, sometimes emotionally.

On the larger scale, we divide ourselves from one another in all kinds of ways, on the basis of any trait that distinguishes one human being from the next. Here are the main areas to which I believe humanity could choose to bring a spirit of reconciliation:

- **Economic reconciliation**—Disparities between the rich and the poor are enormous, and they keep growing. Reconciliation would require narrowing these differences and establishing a minimum standard of economic well-being for all people. Economic reconciliation also suggests that wealthier individuals and nations would begin to voluntarily simplify the material side of life and shift increasing attention into psychological, cultural, and spiritual growth and to assist those living in extreme poverty.

- **Racial, ethnic, and gender reconciliation**—Discrimination on the basis of race, ethnicity, gender, and sexual orientation profoundly divides humanity. How can we create a common future unless we can develop mutual respect? The healing of relations between different groups will transform the psychic wounds of humanity's history.

- **Spiritual reconciliation**—Religious intolerance has produced some of the bloodiest wars in history. Reconciliation among the world's spiritual traditions is vital to humanity's future. It is possible for us to learn to appreciate the core insights of each tradition and to see each as a different facet of the larger jewel of human spiritual wisdom.

- **Generational reconciliation**—Sustainable development has been described as development that meets our needs in the present without compromising the ability of future generations to meet their needs.[10] At present, industrial nations are consuming nonrenewable resources at a rate that will handicap future generations. We have the opportunity to reconcile ourselves with generations yet unborn. We would be wise to use as our example the Iroquois, who, in making major decisions look at the expected impact seven generations ahead.

- **Species reconciliation**—Living in sacred harmony with the Earth is essential if we are to survive and evolve as a species. Our future depends on the integrity of our ecological system, whose strength depends on a broad diversity of plants and animals. We have the opportunity to reconcile ourselves with the larger community of life on

Earth. To do so would be to move from indifference and exploitation to reverential stewardship.

Although there is continuing conflict in each of these areas, there is also new hope for reconciliation. Let's look briefly at changes happening in just one area—the relationship between women and men, which is shifting from patriarchy to partnership. Global surveys show that the status of women is improving and that their acceptance in a partnership role in society is increasing. In 1995, the Gallup organization conducted the *Gender and Society* survey in twenty-two countries in Asia, Europe, North America, and Latin America.[11] The survey found that in most countries, large majorities said that job opportunities should be equal for men and women. In all countries but one, majorities believed that their country would be governed better if more women were involved in politics.

Because the relationship between men and women is so basic, a shift in gender relations from domination to partnership dramatically increases the possibility that we will live more cooperatively and sustainably. Susan Davis, an activist who has worked internationally for both gender equality and sustainability, concludes that equality is not a luxury but is a prerequisite for sustainable development: "We're talking not just about ending oppression. We're talking about unleashing leadership, creativity, and real wisdom. We will not get there without achieving gender equality."[12]

We could say much the same thing about any group that has been oppressed. To have a sustainable future, we need the creativity and talents of the whole human family. Without reconciliation, that will not be possible.

It is important to emphasize that reconciliation does not mean homogeneity. Quite the contrary, our diversity is fundamental to our success. According to the historian Arnold Toynbee, homogeneity appears to weaken societies, while diversity appears to strengthen them. After surveying the growth and decline of the world's major civilizations, he found that a "tendency toward standardization and uniformity" marked their disintegration while the opposite, a "tendency toward differentiation and diversity," marked their growth.[13] As we seek reconciliation, it is important for us to preserve and learn about the unique gifts of culture and history of different people and groups.

The Process of Reconciliation

RECONCILIATION does not mean forgetting the suffering and injustices of the past; rather, it means not letting the past stand in the way of opportunities for the future. When historic injustices are publicly acknowledged and realistic remedies are found, hurts from the past no longer stand in the way of collective progress. Freed from the need to continue blaming and feeling resentful, people can shift their focus from past grievances to mutual opportunities in the present and the future.

The process of reconciliation is complex and involves at least three steps: the injured need to be heard publicly, the wrongdoers need to apologize publicly, and if appropriate, they need to provide restitution or reparations.

Being heard is the first step in being healed. By listening to and acknowledging the stories of those who have suffered, we begin the process of healing. Our collective listening to the wounds of humanity's psyche and soul is vital to our collective healing.

In *An Ethic for Enemies,* his book on the politics of forgiveness, Donald Shriver, Jr., explains that in popular usage, the phrase "to forgive" is thought to mean "to forget." But, he says, that is not what forgiveness means: "Instead, 'remember and forgive' would be a more accurate slogan."[14] Forgiveness requires that we surrender revenge as a basis for justice. We need to call forth mercy and forbearance in order to break through the cycle of violence and counterviolence. Forgiveness also requires that the injured seek to understand the actions of the wrongdoer so as to restore the wrongdoer's humanity. As a final step in reconciliation, both parties need to create a new relationship so that they can live together in peace and mutual respect.

Archbishop Desmond Tutu knows more about the process of reconciliation than most of us do. He was the chairman of the Truth and Reconciliation Commission (TRC), which was established to investigate crimes committed during the apartheid era in South Africa from 1960 to 1994. He describes the logic of reconciliation in his country in this way. When apartheid ended, South Africa's black majority had to choose among three ways to seek justice and continue to live together with the country's white minority. They could have chosen justice based on *retribution*—an eye for an eye; on *forgetting*—don't think about the past, just move forward into the future; or on *restoration*—granting amnesty in exchange for truth. This is how Tutu explains their choice:

> We believe in restorative justice. In South Africa, we are trying to find our way toward healing and the restoration of

> harmony within our communities. If retributive justice is all you seek through the letter of the law, you are history. You will never know stability. You need something beyond reprisal. You need forgiveness.[15]

The commission received over seven thousand applications for amnesty in exchange for truthful accounts of violations of human rights. Before the commission finished its work in 1998, nearly two thousand people testified before it, and it received roughly twenty thousand statements of rights abuses. In concluding the work of the TRC, Tutu said that although he was "devastated by the depths of depravity that the process has revealed," he also had "been amazed, indeed exhilarated by the magnanimity and nobility of spirit of those who, instead of being embittered and vengeful, have been willing to forgive those who treated them so horribly badly."[16] The deputy chairman, Dr. Alex Boraine, said that perhaps the greatest contribution of the TRC toward social reconciliation was the recognition that "reconciliation is not easy, is never cheap and is a constant challenge." Although the process of bringing closure to the era of apartheid was messy and agonizing, it was effective in creating the foundation for a new beginning. Boraine explained that many people testified that appearing before the commission finally put an end to their "nightmares of isolation" and that, for the first time since losing their loved ones, they could sleep at night. Yet others told of a "broken heart which had been healed."[17]

The strong sense of community found in South African culture helped to inspire this approach to reconciliation. In the African view, the community defines the person. The word for this is *ubuntu,* which, translated roughly, means "each individual's humanity is ideally expressed in relationship with others" or "a person depends on other people to be a person."[18] From this feeling for community emerged the nonviolent means to move South Africa from racial separation and minority rule to integration and democracy.

A second step in the process of reconciliation is for the wrongdoer to offer a sincere public apology. Here are examples of important public apologies offered in recent years:[19]

- In 1988, an act of Congress apologized "on behalf of the people of the United States" for the internment of Japanese Americans during World War II.
- In 1996, German officials apologized for the invasion of Czechoslovakia in 1938 and established a fund for the reparation of Czech victims of Nazi abuses.
- In 1998, the Japanese prime minister expressed "deep remorse" for Japan's treatment of British prisoners during World War II.

A powerful example of a public apology and social healing is provided by the relationship between the Aboriginal people and the European settlers in Australia. On May 26, 1998, Australia commemorated its first "Sorry Day" to express people's regret and shared grief about a tragic episode in Australian history—the organized removal of Aboriginal children from their families on the basis of race. Through much of this century, Aboriginal children were forcibly removed from their families with the aim of assimilating them into Western culture.[20] According to an indigenous council member, Patricia Thompson, Sorry Day provides a way for Australians to come to terms with their history and to come together to build a future on a foundation of mutual respect. Said Thompson, "What we want is recognition, understanding, respect and tolerance—of each other, by each other, for each other." In cities, towns, and rural centers, in schools and churches, people stopped their everyday activities to acknowledge this injustice. In addition, hundreds of thousands of Australians have signed the "Sorry Books."

The third step in reconciliation is restitution or the payment of reparations. Archbishop Desmond Tutu gives a good explanation of the role of restitution when he says that completing the process of reconciliation involves more than the recognition and remembering of injustice: "If you steal my pen and say 'I'm sorry' without returning the pen, your apology means nothing."[21] In cases like this, what is needed is restitution. Apologies create a truthful record. Restitution creates a new record. The purpose of reparation is to repair the material conditions of a group so as to restore some balance or equality of power and material opportunity.[22]

Beyond reconciliation is the day-to-day reality of former antagonists living together. One of the most notable examples of successful reconciliation within the recent past is the shift in the relationship between the United States and Germany and Japan. World War II began in the age of total warfare, when massive civilian casualties were the norm; and it could have taken many generations to heal the psychological wounds from that war. Yet, within a few decades, the United States and its bitter enemies from the war, Germany and Japan, became peaceful allies—clear examples of successful reconciliation culminating in renewed relationships and mutual respect. Other important examples of reconciliation include the ongoing peace process in the Middle East, certainly one of the most volatile regions in the world, and in Northern Ireland, where the process of reconciliation seems poised to overcome centuries of separation and conflict.

As these examples make clear, with authentic reconciliation—with listening, apologizing, and restoring—the suffering of the past does not need to stand in the way of future progress.

The Cost of Kindness

THE cost of compassion is far less than we might think. The world does have the material resources for all of us to live together sustainably. We could begin by eliminating the worst aspects of poverty—a fundamental requirement, I believe, for an evolutionary bounce to occur. As the *United Nations 1998 Human Development Report* concludes, we have "more than enough" resources to accomplish this.[23] To make this point, the report presents these stark contrasts:

- To achieve universal access to water and sanitation, the estimated additional annual cost is $12 billion, which is what is spent on perfumes in Europe and the United States each year.
- To achieve universal basic health and nutrition, the estimated additional annual cost is $13 billion, which is $4 billion less than annual expenditures on pet foods in Europe and the United States.
- The world's spending priorities are further reflected in these figures: Annual expenditures on business entertainment in Japan amount to $35 billion; on cigarettes in Eu-

rope, $50 billion; on alcoholic drinks in Europe, $105 billion; and on military spending in the world, $780 billion.

The *Human Development Report* concludes that "advancing human development is not an exorbitant undertaking." The added bill to provide universal access to basic services—education, health, nutrition, reproductive health, family planning, safe water, and sanitation—is estimated to be an additional $40 billion per year.[24] This is less than one-tenth of one percent of world income. As the report notes, this is "barely more than a rounding error."

Given that we can easily afford to eliminate the worst forms of poverty, what are we doing about it? The report states that development aid is now at its lowest level since the U.N. started keeping statistics. Donor countries allocate an average of only 0.25 percent (one-quarter of one percent) of their total GNP to development assistance for poorer nations. The United States is the stingiest developed nation in terms of the proportion of total wealth that it donates.[25]

The resources exist to make a dramatic improvement in the quality of life for a majority of humanity and to begin a process of reconciliation between the rich and the poor. Instead of trickle-down development from the wealthy to the poor, we could launch a bottom-up approach that directly targets the poor and the voiceless.[26] If we use equity, simplicity, and cooperation as our guideposts, we have the resources to sustain all of humanity into the foreseeable future. As Gandhi said, "We have enough for everyone's need, but not for everyone's greed."

We cannot achieve our maturity if we remain divided into a minority that has great wealth and a majority that is consigned to absolute poverty. We need to create a future of mutually assured development—where progress leaves no one behind and also strengthens the ecosystems on which our common future depends. We could create something akin to the Marshall Plan, which restored Europe after World War II. The entire world could be united in establishing a foundation of sustainability. Given intelligent designs for living lightly and simply, a decent standard and manner of living could vary depending on local customs, ecology, resources, and climate. Within this diversity, if the human family saw its collective development as its central enterprise, the world would have a strong foundation for an evolutionary bounce.

Archbishop Desmond Tutu said that you can immediately tell when you enter a happy home: "You don't have to be told; you don't have to see the happy people who live there. You can feel it in the fabric, the air."[27] In a similar way, he says, we have it in our power to create a cultural atmosphere on Earth that is infused with kindness, joy, laughter, truth, and love. If we can bear witness to the reservoir of unresolved pain that has accumulated through history, we will release an enormous store of pent-up creativity and energy. Instead of mobilizing around enemies, we could release our collective energy in building a future that is worthy of our name as doubly wise humans.

Chapter Seven

Evolutionary Crash or Evolutionary Bounce: Adversity Meets Opportunity

The future enters into us,
in order to transform itself in us,
long before it happens.
—*Rainer Maria Rilke*

Even if you are on the right track,
you'll get run over if you just sit there.
—*Will Rogers*

The Dynamics of Initiation

THE preceding chapters explored the titanic forces converging at this time in human history. Powerful adversity trends—global climate change, the rapid extinction of species, the depletion of key resources such as water and cheap oil, a burgeoning population, and a growing gap between the rich and poor—are converging into a whole-systems crisis, creating the possibility of an evolutionary crash. At the same time, four equally powerful trends are converging into a whole-systems opportunity, creating the possibility of an evolutionary bounce. The first opportunity trend is the emergence of a new perceptual paradigm that invites us to see the aliveness and unity of the universe. In seeing through the deadness of materialism and into the subtle aliveness that infuses the world around us, we transform the human experience from a secular to a sacred journey. The second trend is the shift toward simpler ways of living. With the ability to voluntarily simplify our lives, we can choose ways of life that are more satisfying and more sustainable. The third trend is the global communications revolution, which gives us the tools to communicate our way into a positive future. We can now engage in conversations about our common future both locally (such as across the kitchen table) and globally (such as via the internet), recognizing that the very act of doing so has thc power to ripple out into a world hungry for visions of hope. These three trends make possible the fourth trend, that of reconciliation—between men and women, black and white, rich and poor, humans and other species, current and future generations, and many more. Reconciliation gives us the ability to heal old wounds that keep human beings from unleashing our potentials.

From the meeting of adversity trends and opportunity trends comes our time of initiation into our adulthood as a human family. There is a profound irony in all this: we are initiating ourselves. By our own actions, we are creating the circumstances that are now challenging us to reach new heights of maturity and community.

As you may recall from chapter one, the purpose of an initiation is to enable a person or community to cross a threshold from one life into another, and to mark the transition into a new level of social membership. Historically, the process of initiation is characterized by shared ordeals, adventures, and discoveries that produce bonding and a sense of a new beginning that overcomes old differences. That is precisely the opportunity I see for humanity—to cross the threshold from our adolescence into our adulthood with a sense of authentic membership in the human family.

Another way of looking at our coming time of initiation is to see it as an important experience in the awakening of our species as *Homo sapiens sapiens*. Recall that humans are regarded as more than *sapient* or "wise"; we are *sapient sapient* or "doubly wise."[1] Whereas animals "know," humans "know that we know." In that subtle circling back of consciousness upon itself, a revolutionary potential is born. In his book *The Phenomenon of Man,* Teilhard de Chardin says that when the first living creature consciously "perceived itself in its own mirror, the whole world took a pace forward."[2] The capacity for self-observation or double wisdom is not a trivial enhancement of evolutionary potential. It is an explosively powerful capacity that has given a supercharged boost to the evolutionary process. As Teilhard de Chardin says, "The being who is the object of his own reflection, in consequence of that very doubling back upon himself, becomes in a flash able to raise himself into a new sphere."[3]

When a group—a tribe, a nation, or an entire species—"knows that it knows," it has the ability to be self-observing and to take responsibility for its actions. For the first time in our history, humanity is acquiring the means to "know that we know" as a species and, just as this capacity is explosively powerful for individuals, so too can it give a supercharged boost to humanity's evolution. When humanity consciously recognizes itself as a single community with responsibilities to the rest of life, both present and future, we will cross the threshold to a new level of maturity and a new culture and consciousness will begin to grow in the world.

How soon might the human family move into a time of initiation? In my view, although many of the adversity trends are already quite serious at the turn of the millennium, they are still not sufficiently compelling to awaken humanity's collective attention and galvanize us into concerted action. There seems to be enough resilience left in the world system to absorb further economic, environmental, and social shocks without being pushed to the breaking point. But while the problems that I am calling "adversity trends" may seem manageable and relatively independent of one another at the turn of the millennium, I believe they will soon coalesce into a tight and unyielding web, most likely by the decade of the 2020s. My guess is that, when this happens, our planetary system will be like a rubber band stretched to its limits: it will lose much of its elasticity and ability to cope with further disruptions. This is a condition ripe for either breakdown or breakthrough.

Our coming initiation will be a defining period in human history. The choices we make at this decisive time will reverberate far into the future. After traveling a path of separation and division for the past thirty-five thousand years, humanity will be challenged in a brief moment of time to make a conscious turn toward a new life of connection and cooperation. Will we make a successful turn toward our higher maturity? As we rush headlong toward the answer to that question, let us look briefly at the major stages through which we seem likely to move in the next few decades:

Stage I: Denial—Many insist that things will not become so chaotic and turbulent as to require major change. Others say that great changes have happened through history, so there is nothing unique about this time, and no particular reason to be concerned. Others consider the situation far too complex and feel powerless to get involved. Still others recognize the urgency of change but procrastinate, not wanting to upset a comfortable (or at least predictable) way of life.

Stage II: Innovation—Without public fanfare, a growing number of people are crossing a threshold of awakening and launching innumerable social and technological innovations. While the larger society may be relatively unaware of what is occurring, there is intense grassroots activity under way to design ourselves into a workable and meaningful future. Diverse networks of communication—ranging from small local groups to internet-based global dialogues—are vital to this burst of innovation.

Stage III: Initiation—By the time the adversity trends and the opportunity trends reach a critical point of convergence, only one outcome seems certain: humanity will enter the fire of initiation. We will move into a time of intense, planetary compression. A circle will have closed. All frontiers will be gone. There will be nowhere to escape. It will be obvious that the Earth is an integrated system—environmentally, economically, and socially. The forces of adversity will intensify one another and become so unyielding, and the forces of opportunity will become so promising, that humanity will be pushed to make an unprecedented choice. Do we hold on to the past and descend into chaos, or liberate ourselves from historical limitations so that we can consciously create our future? I believe that we will not be able to stop the process leading to initiation any more than a pregnant woman can stop the birth of her baby. With no way back, our success lies in consciously moving forward.

Stage IV: Bounce or Crash—The outcome of our journey is yet to be determined. Will we reconcile ourselves around building a future worthy of our potentials? Or will humanity be torn apart by adolescent conflicts as we squander our abundant resources and mutilate our home?

Figure 6 presents a highly simplified view of the dynamics likely to be involved if humanity moves successfully through our evolutionary initiation in the coming decades.

FIGURE 6

Stages of Evolutionary Initiation and Transition

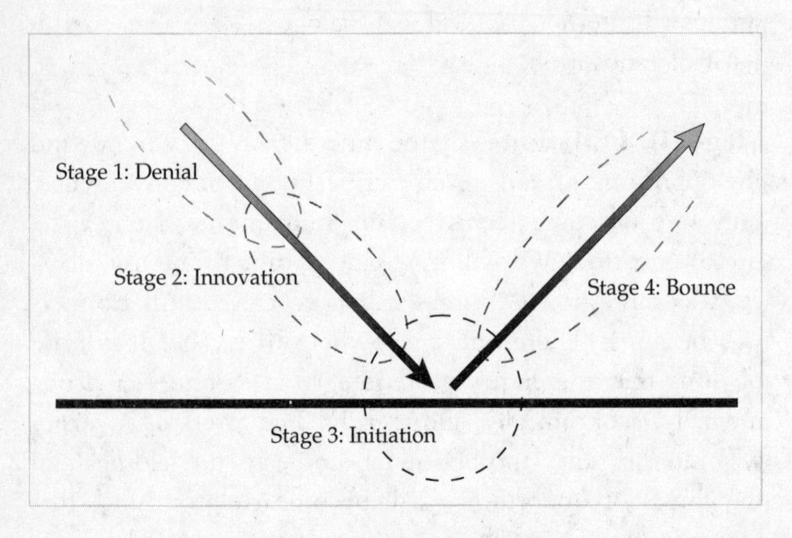

If this sequence of stages is roughly accurate, then the time of greatest opportunity for social reflection and creative invention is prior to our entry into the stage of initiation with its demands for immediate action. Stated more bluntly, it takes little collective intelligence to crash blindly into an evolutionary wall and then decide it would be wise to take corrective measures. If that is the best we can muster, then our future is in great peril. By my reckoning, we have two precious decades between the turn of the millennium and the 2020s before a global systems challenge seems likely to emerge full force. A fundamental test of our maturity as a species will be how well we use this brief window of opportunity for genuine social dialogue and innovation in support of a sustainable and meaningful future.

Ultimately, I believe that whether we experience an evolutionary bounce or an evolutionary crash will be determined by whether the human family chooses to pull together and cooperate, or to pull apart and compete. To pull together implies that we are on a shared journey, and that we have a common story and a common purpose. It means that our social and cultural fabric is resilient, flexible, and able to stretch without tearing. If we pull together, we can conserve the evolutionary momentum of the past and use it to carry us into a promising future. If, instead of pulling together, we struggle and separate—rich against poor, men against women, black against white, nation against nation—we may do lasting harm to the fabric of human connection, and veer off on an evolutionary detour.

A crash or a bounce may seem like a stark set of choices, so let's pause to consider other outcomes. Are these the only two scenarios that are possible? Not necessarily. It is possible that, with a combination of high technology and good luck, the human family could find a way to mature without undue hardship. Although this is a happy prospect, it is one that I find highly unlikely for all of the reasons outlined in chapter two. Because adversity trends seem to be unyielding, I think a more likely outcome is one that combines elements of both the crash and the bounce—a condition described by historian Arnold Toynbee as dynamic stagnation or "arrested growth."[4] Arrested growth occurs when a civilization is unable to advance and yet has gone too far to turn back, when all of the society's energy is being used just to stay where it is. A future of dynamic stagnation would mean that we would have to run as fast as we could simply to maintain a stalemate between the forces of advance and the forces of collapse. By intensifying efforts that worked well in the past, we may be able to hold on, maintaining ourselves but unable to muster the energy and creativity to surpass ourselves, while becoming progressively weaker. This is a delaying tactic, not a long-term solution and the eventual outcome is still likely to be either an evolutionary crash or a bounce. Therefore, I want to focus on the heart of the evolutionary choice: Do we strive for a new level of maturity and wholeness as a human family (an evolutionary bounce), or do we pull apart in conflict (an evolutionary crash)? This chapter explores these two possible outcomes.

An Evolutionary Crash

I do not view the extinction of humanity as the likely outcome of an evolutionary crash. Because I expect humans to have an enduring presence on the Earth, my concern is not with our extinction but with our evolution. An "evolutionary crash" refers to a breakdown in our capacity to unfold our potentials, both personal and collective. It means that when the human family confronts monumental adversity (primarily of our own making), we will not able to move beyond our adolescent consciousness and behaviors. Instead of pulling together for our common future, we may pull even further apart. If this happens, the result could be a deep wounding of the biosphere as well as of humanity's collective psyche. We may find our collective psyche so wounded and crippled by the experience of initiation that our ability to evolve to a higher maturity would be diminished for generations to come. Instead of surpassing ourselves, we would be struggling just to hold on—to maintain ourselves. Living would become little more than "only not dying," to use Simone de Beauvoir's phrase. While I am confident that humanity will survive, I am apprehensive about the kind of life that we may create for ourselves together.

Even if we should experience some form of evolutionary crash, we may have more chances to evolve and grow in the future. What we would miss is an unusually favorable opportunity to advance—one that may not reappear for generations, if it at all.

Although history has seen the rise and fall of more than twenty major civilizations, what happened on Easter Island provides the most vivid—and chilling—example of how devastating an evolutionary breakdown can be. It illustrates what could lie in humanity's future if we do not rise to a higher level of maturity and halt the "irretrievable mutilation" of the natural environment.

The story of Easter Island. Located in one of the most remote places on Earth—in the Pacific Ocean, roughly two thousand miles off the coast of South America—Easter Island is only one hundred fifty square miles in area. A person can walk around it in about a day. The first Europeans to visit the island were the crew of a Dutch ship that arrived on

Easter Sunday in 1722—hence the name Easter Island. They found a primitive society of approximately three thousand people, living in wretched reed huts and caves, engaged in almost perpetual warfare, and resorting to cannibalism in a desperate attempt to supplement the meager food supplies available on the treeless island.[5] What was most amazing to the Dutch were the massive stone statues that covered the island. There were more than six hundred, each averaging over twenty feet in height, with some as tall as forty feet. The Europeans could scarcely believe that the primitive and impoverished islanders were capable of carving, transporting, and erecting so many enormous statues, whose presence indicated that an advanced society had once flourished there.

Archeological evidence reveals that when Easter Island was first settled by a few dozen Polynesian colonists in approximately A.D. 500, it had a mild climate and volcanic soil, was covered by forests, and teemed with animal and plant life (although, given the island's remoteness, there were relatively few species). Among the foods that the settlers brought with them, yams and chickens were particularly suited to the climate and soil. Although the diet was monotonous, the islanders prospered, and their numbers grew to an estimated seven thousand when the population peaked around 1550.[6]

Because food production was so easy, the islanders had abundant free time. Families coalesced into clans that competed with one another in creating elaborate rituals and erecting the massive statues. Over a thousand years, they developed one of the most advanced and complex societies in the world, despite their limited resources and technologies.[7] From early on, however, they used the resources of the island beyond their regenerative capacity. Archeological evidence shows that the destruction of the island's forests was well under way by the year 800—only three hundred years after settlers first arrived. By the 1500s, the forests and palm trees had disappeared as people cleared land for agriculture and used the surviving trees to build homes and oceangoing canoes, burn as firewood, and transport statues. The islanders apparently used logs to move and raise the statues in a competitive rivalry to see which clans could build the most. At the end, the remaining forests disappeared quickly. It is astonishing to note that given Easter Island's small size, everyone would have been able to see when the last tree was being cut down, thereby sealing their fate as prisoners on the tiny island. The loss of tree cover increased soil erosion and reduced soil quality, and thus crop yields declined.

The ecological destruction was not confined to the forests. Jared Diamond, professor of medicine at UCLA, describes how the animal life was also eradicated:

> The destruction of the island's animals was as extreme as that of the forests: without exception, every species of native land bird became extinct. Even shellfish were overexploited, until people had to settle for small sea snails . . . Porpoise bones disappeared abruptly from the garbage heaps around 1500; no one could harpoon porpoises anymore, since the trees used for constructing the big seagoing canoes no longer existed . . .[8]

By the mid 1500s, the biosphere was so devastated that it was beyond short-term recovery. With no trees to build boats, ocean fishing was impossible. With animals hunted to extinction, the people turned on one another. Centralized authority broke down, and the island descended into chaos. Rival clans lived in caves and competed with one another for survival. Eventually, according to Diamond, the islanders "turned to the largest remaining meat source available: humans, whose bones became common in late Easter Island garbage heaps. Oral traditions of the islanders are rife with cannibalism."[9] Warfare continued after the Europeans left and, by the late 1700s, the population had crashed to between one-quarter and one-tenth of its peak level. Here is how author Clive Ponting summarizes the rise and fall of civilization on Easter Island:

> Against great odds the islanders painstakingly constructed, over many centuries, one of the most advanced societies of its type in the world. For a thousand years they sustained a way of life in accordance with an elaborate set of social and religious customs that enabled them not only to survive but to flourish . . . But in the end the increasing numbers and cultural ambitions of the islanders proved too great for the limited resources available to them. When the environment was ruined by the pressure, the society very quickly collapsed with it, leading to a state of near barbarism.[10]

The parallels between Easter Island and the Earth are strong. Professor Diamond concludes, "Easter Island is Earth writ small. Today, again, a rising population confronts shrinking resources . . . we can no more escape into space than the Easter Islanders could flee into the ocean."[11] As Easter Island reveals, we humans have already demonstrated our ability, on a small scale, to descend from greatness into collective madness and devastate an entire biosphere and culture irreparably.

Humanity's potential for collective madness. A concern for the sanity of nations is not new. In 1841, Charles Mackay wrote of the madness of crowds and nations: "in reading the history of nations, we find that, like individuals, they have their whims and their peculiarities; their seasons of excitement and recklessness, when they care not what they do."[12] Nearly a century later, in 1930, Sigmund Freud expressed his concern for the neuroses of civilizations in his book *Civilization and Its Discontents*:

> If the development of civilization has such a far-reaching similarity to the development of the individual and if it employs the same methods, may we not be justified in reaching the diagnosis that, under the influence of cultural urges, some civilizations or some epochs of civilization—possibly the whole of mankind—have become "neurotic"?[13]

I do not think that Freud goes far enough in his assessment of collective madness. He speaks of neuroses—which are relatively mild forms of dysfunctionality—to describe civilizational behavior. This is far too gentle a term for some of the examples of collective madness that have emerged in recent centuries. Let us look at just one of these examples—the witch hunts in Europe during the Middle Ages—as they are a dramatic illustration of humanity's capacity for collective madness.

The witch hunts can be described as nothing less than a period of collective psychosis. During these dark centuries, the idea persisted in Europe that disembodied spirits inhabited the Earth, and that some people—witches—had the power to summon evil spirits so as to bring misfortune to others. As a result, an epidemic of terror seized Europe. Few thought themselves secure from the powers of evil spirits. A witch was suspected as the cause of every calamity. If a storm blew down a barn—it was witchcraft. If cattle died unexpectedly—it was witchcraft. If a loved one suddenly became ill—it was witchcraft. Spurred by invisible terror, people hunted down, tortured, and killed those they thought to be witches.

The witch hunting craze began in earnest in the 1400s with the encouragement of the Catholic Church. A declaration issued in 1484 by Pope Innocent VIII provided the moral authority and official encouragement for the witch hunts. It reads in part:

> . . . many persons of both sexes, heedless of their own salvation and forsaking the catholic faith, give themselves over to devils male and female, and by their incantations, charms, and conjurings, and by other abominable superstitions and sortileges, offences, crimes, and misdeeds, ruin and cause to perish the offspring of women, the foal of animals, the products of the earth, the grapes of vines, and the fruits of trees . . . We therefore . . . remove all impediments by which in any way the said inquisitors are hindered in the exercise of their office . . . it shall be permitted to the said inquisitors . . . to proceed to the correction, imprisonment, and punishment of the aforesaid persons . . .[14]

This was religious madness. The witch hunts resulted in the public torture and cruel deaths of at least several hundred thousand women (as well as many men and children) over a period of two and a half centuries. "Witch mania" generated so many trials for witchcraft in France, Italy, Germany, Scotland, and other countries that for years other crimes were seldom considered. One bishop (in Geneva) burned five hundred "witches" within three months, another bishop six hundred, and another nine hundred.[15] After two hundred and fifty years, this wave of cultural madness began to subside, gradually giving way to the rationalism of science and the Industrial Revolution.

The examples of Easter Island and the witch hunts in Europe show how vulnerable we have been to collective madness. Are we still? I and many others believe that we are. Consciousness researcher Dean Radin conjectures that ". . . there may be a mental analogy to environmental ecology—something like an ecology of thought that invisibly interweaves through the fabric of our society. This suggests that disruptive, scattered, or violent thoughts may pollute the social fabric . . . Perhaps periods of widespread madness, such as wars, are indicators of mass-mind infections."[16]

In the coming planetary initiation, it seems very plausible that the world could go mad with such divergent views and paradigms that we would be unable to come together in meaningful dialogue or to build a working consensus for the future. The conversation of the planet could collapse to the lowest common denominator consistent with security and survival. In the face of monumental stress, and with no overarching and trusted source of perspective, humanity's collective psyche could fragment, and the people of the Earth could descend into perpetual conflict. The conviction could grow that humanity is an ill-fated species that never had a chance to succeed. A further generation could reconfirm the suspicion that we live in a hostile universe, that we do not share a coherent view of reality, that humanity cannot work together, and that we are a doomed species. Particularly in developed nations, people could feel enormous guilt and responsibility for the devastation of the planet and the wasting away of opportunities for future generations. Many could feel that after tens of thousands of years of slow development, we ruined our chance at evolutionary success within the span of a generation or two—and that a new Dark Age now looms ahead for the people of the Earth.

If humanity misses this chance for an evolutionary leap forward, we may have other opportunities. But what we will have allowed to pass by is a moment unique in its promise for realizing our collective maturity. If we do not rise to the opportunity this time, it seems very likely that we will so devastate the biosphere and so burden ourselves with hatreds and resentments that it could take infinitely more effort to realize our evolutionary potentials in the future. We may pay a heavy price if we forgo this evolutionary opportunity.

An Evolutionary Bounce

AN evolutionary bounce represents a period of rapid acceleration in the evolution of the species. However, a leap forward will not happen automatically. The most essential ingredient for creating a bounce will be conscious choice. Instead of flying off in all directions, we can choose to focus our energies and conserve the historical momentum of human development.

The difference between a crash and a bounce will be not so much in the circumstances that we encounter, but rather in how we respond to those circumstances. In a crash scenario, there will be destructive conflict, paralyzing chaos, and deep separation. In a bounce scenario, there will also be conflict, chaos, and separation—but the conflict will be constructive, the chaos will be creative, and separation will be balanced with integration. Barring an environmental catastrophe, all the forces that would be present in a crash would be present in a bounce scenario, but in the latter we would respond to those forces in a more reflective, mature, and productive way.

Although adversity trends are providing humanity with a wake-up call, I see no advantage in allowing them to grow into an irreversible calamity. The destruction of our biosphere is not likely to bring us to collective sanity. On the contrary, if we allow an environmental crash to devastate the Earth, it could be infinitely more difficult to regain our evolutionary direction and momentum. There is already sufficient suffering in the world to motivate change. What is needed is the unflinching maturity to face the coming challenges head on. We need to start acting like adults. When we do, when we consciously recognize ourselves as a single community of life with responsibilities to future generations, we will be taking a leap forward in our evolution.

Given the separation and conflict that have marked our species in the past, it might be difficult to imagine humanity pulling together to create a sustainable and meaningful future. But our pulling together is not a far-fetched idea. There are already many examples of successful, planetary-wide cooperation:

- The world weather system merges information from more than one hundred countries every day to provide weather information globally.

- Nations around the globe have cooperated to eradicate diseases such as smallpox, polio, and diphtheria.

- International civil aviation agreements assure the smooth functioning of global air transport.
- The International Telecommunications Union (ITU) allocates the planetary electromagnetic spectrum so that television signals, cellular phones, and radio signals are not overwhelmed with noise.
- In 1990, the world's nations agreed to ban CFCs—a chemical used in refrigerators and cooling systems that damages the ozone layer.

There is no reason why we cannot expand on these areas of cooperation in the future. We have already begun building the foundation for an evolutionary jump.

I believe that we have all the material resources and technologies we need for an evolutionary bounce. All that we need is the social will to make it happen. Just how great an evolutionary jump we could achieve is vividly demonstrated by the following story of a village that has risen from a grassy desert in South America—Gaviotas.

The story of Gaviotas. The village of Gaviotas is located on the vast, desolate plains of eastern Colombia where nothing but a few nutrient-poor grasses grow. It is surely one of our planet's least desirable places in which to live. Paolo Lugari, who founded the village in 1971, explained why the villagers chose this site: "They always put social experiments in the easiest, most fertile places. We wanted the hardest place. We figured if we could do it here, we could do it anywhere."[17] When people would tell him that the area was "just a big, wet desert," Lugari would reply, "The only deserts are deserts of the imagination."[18] In the space of a single generation—less than thirty years—Gaviotans have transformed one of the most resource-starved regions in the country into a sustainable economy, a nurturing community, and a flourishing ecosystem. In doing so, they have given us a brilliant example of just how rich and fertile the human imagination can be.

In the early 1970s, Lugari brought scientists, engineers, doctors, university students, and other advisors to this remote and inhospitable site to explore how it could be transformed into a thriving community. They produced a dazzling array of low-cost but highly efficient technologies. To pump water, they created a lightweight windmill with blades contoured like the wings of an airplane, able to trap the soft equatorial breezes. They attached highly efficient water pumps to seesaws, so that as children play, they pump water for the community. They erected solar water heaters that could catch the diffuse energy of the sun even on the many cloudy days. They placed underground ducts in hillsides to provide natural air conditioning for their hospital. To provide electricity, they put photovoltaic cells on rooftops. And they developed hydroponic gardens to grow some of the village's food.

The transformation of the local ecosystem has been as remarkable as the development of innovative technologies. Since the early 1980s, the Gaviotans have planted roughly two million Caribbean pine trees, the only tree that would grow in the nearly toxic soil. As a result, the village is now home to more than twenty thousand acres of forest. From the trees, the villagers harvest and sell pine resin, which is used in the manufacture of paint, turpentine, and paper. In addition to providing a source of income for the community, the pine forest has brought fresh nutrients to the soil, cooled the ground, and raised the humidity.[19] In turn, these changes have allowed dormant seeds of native trees to sprout and grow. The sheltering pine trees are enabling a diverse, indigenous forest to regenerate itself with surprising speed. As a result, the local populations of deer, anteater, and other animals are growing. The villagers have decided to allow the indigenous forest to choke out the pine forest over the next century, enabling the area to return to its original state as an extension of the Amazon.

The Gaviotans have been equally inventive socially.[20] Everyone earns the same salary, which is above minimum wage. Many of the basics of life are free, including housing, health care, food, and schooling for the children. With no poverty, there has been no need for police or a jail. Government is by consensus and unwritten rules of common sense. Dogs, pesticides, and guns are not allowed. Alcohol use is confined to homes. Laziness is not tolerated. In this community of social invention, people exude happiness. The people of Gaviotas have a sustainable future, a strong community, meaningful work, and a peaceful life.[21]

As the village grows, its creator envisions new satellite villages. "I see enclaves of maybe twenty families, little satellites surrounding Gaviotas, no more than twenty minutes away by bicycle."[22] He envisions "little island communities where people live in productive harmony with nature and technology. And with each other."[23]

In his book *Gaviotas: A Village to Reinvent the World,* Alan Weisman beautifully summarizes the net result of the Gaviotans' efforts:

> Surrounded by a land seen either as empty or plagued with misery, they had forged a way and a peace they believed

> could prosper long after the last drop of the earth's petroleum was burned away. They were so small, but their hope was great enough to brighten the planet turning beneath them no matter how much their fellow humans seemed bent on wrecking it. Against all skeptics and odds, Gaviotas had lighted a path through a magnificent but darkened land, whose sorrows mirrored a beautiful, embattled world.[24]

With the rapid growth of the internet, information about how to create communities such as Gaviotas will soon be accessible to people around the world. The entire planet will become everyone's backyard as we use the internet to scan for social inventions, farming techniques, and energy-production technologies that make sustainable living possible. The Earth will be alive with inventions—both physical and social—that are exquisitely suited to each particular ecosystem, climate, and culture. Gaviotas demonstrates that even in the harshest conditions, we have the ingenuity and cooperative capacity to create a sustainable and meaningful life for ourselves. This small village is a testament that the human imagination is fertile enough to transform virtually any desert.

The potential for collective awakening. Just as we have a potential for collective madness, I believe that humanity has a corresponding potential for collective awakening—and to a much greater degree than we may imagine. In this section, we shall briefly explore the idea of collective consciousness or a group mind, which is the vehicle or context through which our collective awakening could occur.

The idea of collective consciousness has emerged perhaps most clearly in the work of the Catholic priest and mystic

Pierre Teilhard de Chardin, who wrote about the awakening of a collective field that he termed the "noosphere." The noosphere can be thought of as the planetary mind-field or collective consciousness; it is the product of humanity's entire evolutionary experience and expression. This mind-field is enriched as we develop, whether the development is in language, art, music, technologies, society, or any other area of endeavor. In his book *The Future of Man,* Teilhard de Chardin describes the noosphere this way:

> In the passage of time a state of collective human consciousness has been progressively evolved which is inherited by each succeeding generation of conscious individuals, and to which each generation adds something. Sustained, certainly, by the individual person, but at the same time embracing and shaping the successive multitude of individuals, a sort of generalised human superpersonality is visibly in the process of formation on the earth.[25]

In recent years, many different voices have been telling us that humanity's most basic challenge is to awaken our collective consciousness so that we can respond successfully to the social and ecological problems that confront us. Psychiatrist Roger Walsh writes that "the state of the world reflects our state of mind; our collective crises mirror our collective consciousness."[26] Václav Havel, president of Czechoslovakia, stated in a 1990 address to a joint session of the U.S. Congress, "Without a global revolution in the sphere of human consciousness, nothing will change for the better . . . and the catastrophe toward which this world is headed—the ecological, social, demographic, or general breakdown of civilization—will be unavoidable."[27] Philosopher of consciousness Ken Wilber offers a similar view of our Earth's situation: "Gaia's main problems are not industrialization, ozone depletion, overpopulation, or resource depletion. Gaia's main problem is the lack of mutual understanding and mutual agreement in the noosphere about how to proceed with these problems."[28] Author Marianne Williamson describes the contemporary challenge this way: "There is within every person a veiled, oceanic awareness that we are all much bigger than the small-minded personas we normally display. The expansion into this larger self, for the individual and the species, is the meaning of human evolution and the dramatic challenge of this historic time."[29]

Our evolutionary success depends not only on awakening our collective consciousness but also on promoting its health and healing. The tools that will enable this awakening are the internet and related technologies. The internet will soon connect a billion or more people, twenty-four hours a day, seven days a week. My belief is that, with this level of continuous communication—which is unprecedented in human history—humanity will begin the process of collectively "waking up." A new intelligence and awareness could infuse the human family, providing a powerful lift to our evolutionary efforts.

Would our collective awakening enable us to respond at a higher level to the crises of our time? To get a better understanding of what this may mean, let's look at some of what we know about collective consciousness.

Robert Kenny is a management consultant who has explored group consciousness for many years. He has found considerable evidence that group consciousness can foster a high level of synergy among people. Settings where a group consciousness emerges most naturally include highly skilled sports teams; orchestras and jazz ensembles; teams engaged in risky activities requiring a high degree of physical coordination; semi-autonomous, high-performance teams in organizational settings; organizations with a high degree of commitment to a mission; and intentional communities with actively engaged members and a clear sense of spiritual purpose.[30] In these kinds of groups, members feel strongly connected and view themselves as interdependent parts of a larger team or effort. They develop a genuine concern for the well-being of other members and for the productive functioning of the group as a whole. According to Kenny, groups that develop a shared consciousness can perform tasks fluidly, efficiently, cooperatively and in coordination, with minimal communication. He says that an intuitive connection or empathy gives people the ability to anticipate the actions, thoughts, or words of others in the group; as a result, teammates work as a unit rather than as an aggregate of individuals.[31]

Another person who has explored the group mind is Brian Muldoon, who has many years of experience in the area of conflict resolution, and observing what promotes cooperation in groups and organizations. The group mind can emerge, says Muldoon, when people engaged in a common project let go of their personal frame of reference and begin to think on behalf of the whole—which, he says, is not so different from being in love.[32] What draws out the group mind is work on something larger than one's self—a compelling image of collective possibility, a higher potential that can only be realized together.[33] Muldoon's observations lead me to wonder whether humanity can create an evolutionary project compelling enough to draw out and mobilize our group mind.

The importance of the collective mood and atmosphere of a nation has long been appreciated by leaders. Abraham

Lincoln is an example. He said, "With public sentiment nothing can fail; without it, nothing can succeed."

In the United States, in recent decades, there are several powerful examples of collective awakening made possible by the mass media. Television brought the Vietnam war into America's living rooms and gradually awakened the collective consciousness—and aroused the conscience—of the nation. It also brought the nonviolent marches of the civil rights movement into America's collective consciousness and helped to awaken broad recognition of racial oppression. Finally, in airing the voices and views of women calling for liberation, the mass media helped to catalyze the women's movement and its explicit call for consciousness raising.

We have even experienced group consciousness at the planetary level. One example was in 1969 when the human family experienced a few hours of collective self-observation and shared recognition as several billion people around the world paused to watch the first humans walk on the moon. The power of this event was not only in seeing one of our own species on the moon for the first time, but also in the shared awareness that a significant portion of humanity was watching the live telecast and that we were participants in a singular experience.

As these examples illustrate, there is a faculty of knowing at work in our collective lives. Although subtle and easy to overlook, collective consciousness is a natural part of life and presents itself in practical ways.

However we describe it, I believe the co-evolution of culture and consciousness is the core task in our collective future. The question is, in which direction will our collective consciousness develop—toward collective awakening or collective madness?

Chapter Eight

Humanity's Central Project: Becoming Doubly Wise Humans

Once the journey to God is finished,
the infinite journey in God begins.
—*Annamarie Schimmel*

All things arise and pass away,
but the awakened awake forever.
—*The Buddha*

Becoming Fully Human

TO achieve an evolutionary bounce, we will all need to pull together for a common purpose. But what purpose is so compelling? Sustainability alone does not present a very compelling purpose—it offers little more than "only not dying." Is there a higher purpose that could energize and draw the human family together in a common project?

I believe our core purpose as a species lies largely unexplored in the scientific name that we have given to ourselves. As *Homo sapiens sapiens* or "doubly wise humans," we have the capacity to know that we know. *If we use our scientific name as a guide, then our core purpose as a species is to realize—both individually and collectively—our potential for double wisdom.*

We turn, then, to explore our evolutionary purpose as it relates to the development of our capacity for double wisdom. Portions of this chapter draw from my book *Awakening Earth* (which is available in electronic form at www.awakeningearth.org).

What is the nature of "double wisdom?" Human beings not only have the ability "to know" but we also have the ability to turn the mirror of knowing around and to reflect on the miracle of knowing itself. In other words, we can pay attention to how we are paying attention—a process at the heart of meditation. By consciously choosing to live more consciously, we can make a leap forward in our maturity as well as reconnect with the stream of life and the Mother Universe.

Visualizing the Universe's Evolutionary Intention

HOW does humanity's journey of awakening fit within the larger pattern of evolution in the universe? I believe that we need to draw from the depths of nature's wisdom to discover our aligning purpose for moving into the deep future. Our challenge is to discover the evolutionary pathway most in accord with the universe's preexisting intentions. If we fight against nature, we will be fighting against ourselves—and our evolutionary journey will be one of frustration, stalemate, and alienation. If we cooperate with nature, we will be serving our deepest essence—and our journey will be one of satisfaction and learning.

One way to learn about nature's evolutionary direction is by examining the designs expressed in the physical systems of the universe. Wherever we look in the natural world, we see a recurring organizing pattern at work. The basic physical structure of that pattern is the torus, which has the shape of a donut. At every level of the cosmos, we find the characteristic structure and geometry of toroidal forms. *The torus is significant because it is the simplest geometry of a dynamically self-referencing and self-organizing system which has the capacity to keep pulling itself together, to keep itself in existence.* The torus is nature's most common signature of self-organizing systems. Figure 7 shows six different expressions of this easily recognizable form—from the minute topology around a black hole, to the scale of everyday human experience, and then to the scope of a magnetic field around a galaxy.

It is not by chance that at every level of the universe we find self-referencing and self-organizing systems. I believe there is an evolutionary direction and intention being expressed by nature's designs. Self-organizing systems have a number of unique properties that can help us discern the direction of evolution occurring throughout the cosmos:

· **Identity—**A self-organizing system requires a center around which and through which life-energy can flow.

· **Consciousness—**A self-organizing system is self-creating and therefore must be able to reflect upon itself in order to "get hold of itself."

· **Freedom—**To be self-creating, systems must exist within a context of great freedom.

· **Paradoxical nature—**Self-organizing systems are both *static and dynamic* (they are flowing systems that manifest as stable structures). They are both *open and closed* (they are

FIGURE 7

Different Expressions of the Torus—
the Signature of Self-Referencing and Self-Organizing Systems

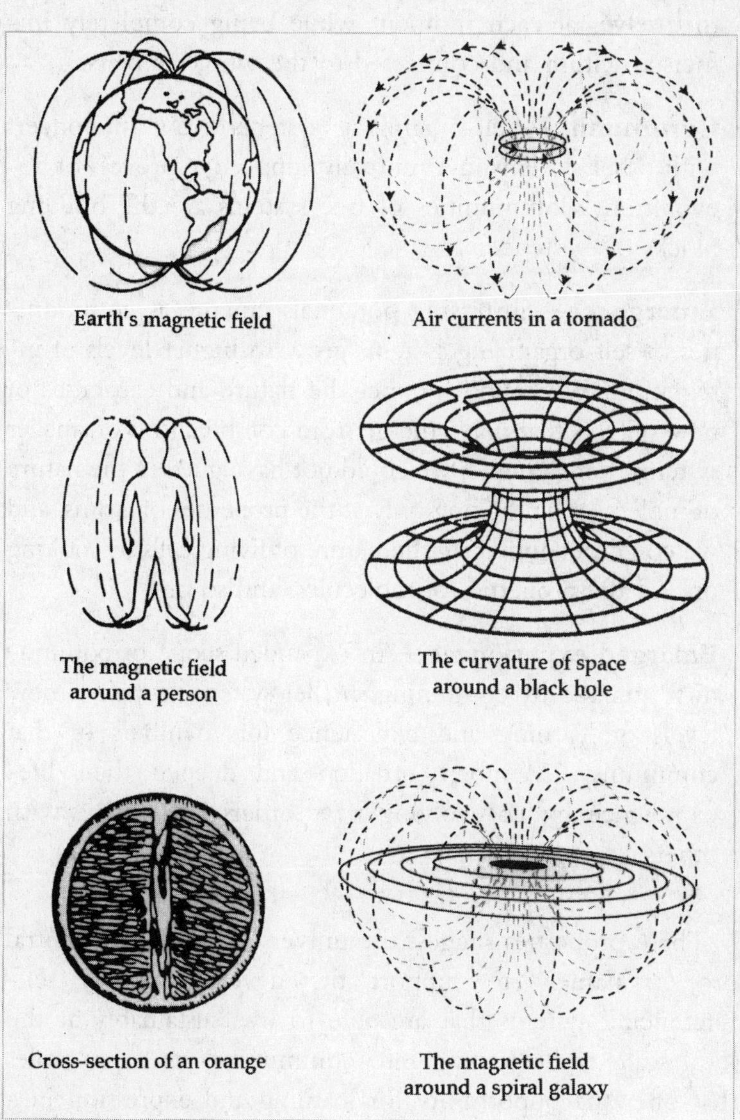

continuously opening to the flow-through of energy and continuously closing into an identifiable entity). And they are both *unique and unified* (they are uniquely manifesting themselves at each moment while being completely immersed within and connected to the whole universe).

- **Community—**Self-organizing systems grow in concert with other systems in a mutually supportive process of co-evolution. Communities of free systems are the building blocks of existence.

- **Emergence—**Surprising potentials emerge as communities of self-organizing systems grow to higher levels of integration. We cannot predict the nature and capacities of new systems that may emerge from combinations of smaller systems. For example, we could not have guessed the nature of molecules by looking only at the properties of atoms, and we could not anticipate the nature of living cells by looking only at the properties of molecules, and so on.

- **Enlarged experiences—**An expanded scope of community, attained by combining smaller systems, supports new levels of learning and experience for members of that community. Members broaden and deepen their life-experience by co-evolving new orders of systems with others.

These properties suggest the universe does have a central project; namely to support the development of self-referencing systems that are able to live sustainably at the local scale while joining into communities at larger scales that offer the opportunity for learning and expression in a context of ever-broadening freedom. Of course, this is no more than an evolutionary orientation. The universe does not provide a cookbook and recipes for unfolding this orientation—and that is the point. Nature is filled with *self*-organizing systems. Evolution is truly a do-it-yourself project.

How does humanity's development into doubly-wise beings fit with the universe's central project of developing self-organizing systems? They are one and the same. If we want to choose an evolutionary direction that is congruent with nature's evolutionary intentions, it will involve developing our species-potential to become fully self-reflective. To the degree that we cultivate this capacity we will, I believe, be successful in our evolution. Becoming fully self-reflective means nothing less than becoming full members of the species *Homo sapiens sapiens*.

As humanity develops its capacity for reflective consciousness, the universe is simultaneously acquiring the ability to look back and reflect upon itself. We are the culmination of 13 billion years of evolution. It has taken the universe these billions of years to grow organisms on the Earth that have a sufficiently developed consciousness to be able to reflect on and appreciate the simple fact of being here, and to be able to look out consciously at the universe in wonder and awe. After billions of years of evolution, a life-form has emerged on the Earth that is literally the Mother Universe looking at her creations through the unique perspective and experience of each person. A gardener appreciating a flower or an astronomer peering out at the night sky represents the closing of a loop of awareness that began with the birth of our cosmos some 12 billion years ago. A Mother Universe gave birth to and sustains a daughter universe, which gave birth to living planets, which, in turn have given rise to life-forms that are now able to know that they know and look back at the universe with wonder and awe. We are returning to where we started and are awakening as doubly wise beings who can consciously reflect on the miracle of the universe and the gift of our existence. In the words of theologian Thomas Berry, humanity enables the universe "to reflect on and to celebrate itself and its deepest mystery in a special mode of conscious self-awareness."[1] The evolution of human consciousness enables the universe to know itself and to experiment consciously in its own evolution. Thus we are partners with the universe in the unfolding story of cosmic evolution.

The Soul's Body

WHY is knowing that we know so central to our life journey? What is so important about this specific capacity that evolution would be centered around it? To me, there is a compelling answer—one that reaches to the limits of human knowledge and language, and then disappears into the great mystery of life and existence.

The world of science and the world of spirituality converge in agreeing that there is a powerful stream of life at the very center of our moment-to-moment experience. *I believe that in developing our capacity for double wisdom, we are directly entering the life-stream and consciously cultivating a body of knowing that is our vehicle through the deep ecology of the Mother Universe—through eternity.* With double wisdom, that stream becomes directly aware of its unfolding existence. From this perspective, a soul is not a static "thing" but a flow of becoming. Each person's life-stream or soul has a unique quality of resonance and texture that is built up through our distinctive experiences and is expressed in our singular personality. In the words of the spiritual teacher Thomas Merton, "Every moment and every event of every man's life on earth plants something in his soul." We live in a world that is exquisitely designed for soul-growing, both for individuals and for entire civilizations. Neither individuals nor civilizations emerge fully evolved. We learn our way into succeeding stages of our maturity. A farmer becomes a farmer by farming season after season; a nurse becomes a nurse by nursing year after year. In a similar way, during the course of our lives we develop the character and texture of our lifestream or soul. In the words of ninety-year-old Florida Scott-Maxwell, "You need only claim the events of your life to make yourself yours. When you possess all you have been and done, you are fierce with reality."

Cultivating the ordinary miracle of our soulful presence in the world is supremely challenging—and uniquely satisfying. A soulful life embraces the full range of living. Just as soul music cuts beneath the surface to deep feelings that are authentic and true, soulful living likewise cuts through to the marrow and truth of life, whether it is heavy with suffering or light on wings of joy.

When we leave this world, none of us will take away a single penny or diploma or title. All of the world's spiritual traditions tell us that, no matter who we are, when we pass on from this world, we take with us only the knowing-resonance or love at the core of our being. That is our true identity. If, during our lives, we have nurtured qualities such as aggressiveness and unrelenting competition, then that is the residue of knowing-resonance that we will take from our brief passage through this world. On the other hand, if the soulful resonance at the center of our being has been nurtured with kindness and enthusiasm for living, then those are the qualities that we will take when we leave this world.

This is not only our learning. Everything we do and learn is woven into the living ecology that is the Mother Universe. We each learn and act on behalf of the totality. In the long sweep of cosmic evolution, everything material will eventually vanish; however, the intensity, character, and texture of knowing-resonance that we experience and nurture within ourselves lives in eternity.

Here are four characteristics of our soulful experience that we can cultivate in everyday life:

- **The soul is a body of love**—Spiritual traditions throughout history have affirmed that we are created from life-energy whose essential nature is love. It is our core essence, at the very heart and center of our experience. Because love is at the center of all, we are able to make friends with ourselves and others. Teilhard de Chardin wrote that the only way for billions of diverse individuals to love one another is "by knowing themselves all to be centered upon a single 'super-center' common to all." The love that is the life-energy of the Mother Universe provides that common center for us all.

- **The soul is a body of light**—The new physics describes light as the basic building block of material reality. In this sense, matter can be viewed as "frozen light."[2] Because we live in a universe of light, it seems fitting to consider the soul as a more subtle body of light that has the potential to evolve into more subtle ecologies of light after the physical body dies. It is interesting that, according to the Gospel of Thomas, when Jesus was asked by a disciple to describe where he came from, he replied "We came from the light, the place where the light came into being on its own accord and established itself."[3] Elsewhere in this same gospel, Jesus says, "Whoever has ears, let him hear. There is light within a man of light, and it lights up the whole world."[4] In Eastern traditions, we find similar descriptions that we are beings of light. In Buddhism, the awakened consciousness is described with phrases such as "self-luminous recognition." It is with good reason that awakening experiences are often described with phrases such as "enlightenment," "seeing the light," and "becoming illumined."

- **The soul is a body of music**—Quantum physics describes all that exists in terms of wave-forms or patterns of resonance. Not surprisingly, we each have a complex hum or resonance that is instantly evident to others as they encounter us and experience our feeling-tone. We are each literally a body of music. Given different temperaments and personalities, each individual is a unique symphony of knowing-resonance.

- **The soul is a body of knowing**—When we allow our ordinary experience of knowing to relax into itself, we find a self-confirming presence. When we know that we know without the need for thoughts to confirm our knowing, we are entering our life-stream directly. As we cultivate our capacity for mindful living or self-referencing knowing, we lessen the need for a material world and physical body to awaken the knowing process to itself. Ultimately, the body that provided the aligning structure to awaken self-referencing knowing can die, and we will endure as a body of light and knowing with the freedom to develop in more subtle ecologies beyond this realm. Once grounded in our capacity for double wisdom, we can be self-remembering in these subtle ecologies without fear of forgetting ourselves. There is no more elevated task than to learn, of our own free will, the skills of living in eternity.

In the view of many spiritual traditions, the soul is a body of love, light, music, and knowing. The qualities we cultivate during our brief stay on Earth infuse the soul, and when our physical body dies, these qualities remain with the soul or life-stream, which is our vehicle through eternity.

If we do not recognize ourselves as a body of love, light, music, and knowing while we live in a physical body, we can overlook ourselves when our physical body dies. Jesus said, "Take heed of the Living One while you are alive, lest you die and seek to see Him and be unable to do so."[5] In Buddhist terms, it is precisely while we have a physical body that we need to recognize our core nature as pure awareness or as the "ground luminosity."[6] In the words of the fifteenth-century Hindu and Sufi master, Kabir:

The idea that the soul will join with the ecstatic
just because the body is rotten—
that is all fantasy.
What is found now is found then.
If you find nothing now,
you will simply end up with an apartment in the City of Death.
If you make love with the divine now,
in the next life you will have the face of satisfied desire.[7]

If we do not use our physical body and world to discover that we already embody the gift of immortality, then when we die we may look out from our subtle body of light and awareness and not recognize our refined existence. If we use our time on Earth to come to self-referencing awareness, we will have anchored the gift of eternity in direct knowing. We can then evolve through the ever more subtle realms of the Mother Universe.[8]

At the beginning of this chapter I questioned what purpose could be so compelling that it could draw the human family together into a common project. What could be more compelling than discovering the priceless gift of eternity? And because life is so interdependent, it is a journey that we are taking together.

The Unfolding of Double Wisdom

ALTHOUGH there have surely been awakened individuals throughout history and in every major culture of the world, my concern is with the changing consciousness of the majority of humans beings—the changing "social average." How has the complex capacity for double wisdom awakened and developed for the overall human family?

Just as there are relatively distinct stages that characterize the development of an individual from infancy to early adulthood, there are also discernible stages in the development of our species. However, to discover those stages, we need to look beneath the people and events that make headlines and that tend to float on the surface of the stream of life. It is the deeper changes, working below the surface of popular culture, that ultimately make history.

In this section I shall explore seven stages of growth that I believe are vital for developing humanity's full potential for double wisdom or reflective consciousness. I recognize that a stages-of-growth description of human evolution can give the impression that evolution is linear—a march from one stage to the next in a smooth and direct flow. Of course, it is not. Human evolution is an untidy process that seldom conforms to orderly progressions and clear boundaries. Our path through the various stages will surely be filled with many surprises, accidents, and confusing twists and turns that will make it uniquely human and characteristically unpredictable. With these qualifications, I do think that there is a general direction to evolution that leads toward our initial maturity as a planetary civilization.

Reflective consciousness is a rich and multifaceted faculty whose full range of potentials develops through a series of stages or learning environments. At each stage, a different set of observing or reflective potentials is awakened, developed, and integrated. Our evolutionary challenge is to consciously retain the lessons of each era while moving on to the next. Our consciousness and culture are maturing through a nested series of experiences, and the complete spectrum is vital in order for us to become fully human.

Stage 1: The era of archaic humans—contracted consciousness. For several million years, our archaic ancestors lived in the faint dawn of reflective consciousness. Their capacity for knowing that we know was almost entirely undeveloped; consequently, our earliest human ancestors functioned primarily on instinct and habit. As a result, their way of life remained virtually unchanged over thousands of generations. Stone tools, for example, show a monotonous sameness over an immense span of time—for roughly ten thousand generations there is no evidence of invention.[9] Some degree of reflective consciousness must have begun to awaken more than a million years ago, when *Homo erectus* migrated out of Africa and, to cope with the harsh Ice Age climate, learned to use animal skins for warm clothing, construct shelters, and use fire. Nonetheless, it is only with the earliest evidence of burials, approximately sixty thousand years ago, that we find a clear recognition of death and evidence of conscious reflection on the "self" that lives and then dies.

Stage 2: The era of awakening gatherers and hunters—sensing consciousness. Although the glowing ember of double wisdom was passed along by our ancestors for several million years, it did not emerge as a distinct flame of self-observation and reflective knowing until roughly thirty-five thousand years ago. At this time, the glacially slow development of culture and consciousness finally achieved a critical mass, and a flow of development began that leads directly to the modern era. Humans made a dramatic leap in their capacity for self-observation, and this is vividly expressed in tremendous changes in toolmaking, painting, and carving as well as in evidence of expanding social and trading networks.

The capacity for fleeting self-recognition that emerged at this time, however, should not be confused with the stabilized "I-sense" that emerges later. There is enormous evolutionary distance between the capacity for momentary self-observation and a steady mirroring capacity that we can consciously mobilize as we move through life. For awakening gatherers and hunters, life was so immediate that, for the most part, it was not contemplated with reflective detachment; instead, things just happened.[10] Much of the time, people operated on automatic—moving through the repetition and routine of a simple, nomadic life. The world was experienced as up close and immediate—a magical place filled with unknown and uncontrollable forces, unexpected miracles, and strange happenings. Nature was a living field without clear boundaries between the natural and supernatural. Daily life was a mixture of unseen forces and unexplained events, for people had neither the concepts nor the perceptual framework to describe rationally how the world worked.

Social organization was on a tribal scale, and individuals felt themselves to be inseparable from the empathic field of their family and tribal group. People's sense of identity came from affiliation with a tribe and from a sense of intimate connection with nature. With few possessions, there was little basis for material differences, or material conflict. Meaning was found in the direct sensing of and engagement with life. A sensing consciousness was bodily based, directly felt, implicit, and tacit. Without an objectified sense of time—without being able to name it or describe its workings—there was little sense of the future; instead, most things happened in the simple, passing present.[11] Every recurring season and event was a unique miracle: the return of springtime after a long winter, the annual migration of animals, the waxing and waning of the moon—all were mysterious wonders.

Stage 3: The era of farming-based civilizations—feeling consciousness. Roughly ten thousand years ago human perception again expanded. People were able to step back even further from unconscious immersion in nature to see how they could tame nature through farming. Combining the gathering of wild grains with seasonal hunting, they made a gradual transition to a settled way of life. Humans made small, incremental improvements in food-raising that, over time, amounted to a revolution in living. Over thousands of years, people learned to pull weeds from wild fields of wheat to increase their yield, to plant seeds around the margins of wild fields to extend the size of the crop, and to protect the fields from grazing animals. From such modest beginnings came one of the most fundamental transformations the world has ever known. The surplus of food that farming produced made possible the eventual rise of large-scale, urban civilizations.

The mind-set of the agricultural era was cyclical, governed by nature—the seasons go round, but the world remains essentially the same. Life was not perceived to be "going anywhere." The vast majority of people lived in small villages and found meaning through belonging to an extended community. No longer were blood and tribal ties the primary cultural glue. In an increasingly differentiated society, it was the power of fellowship, emotional bonds, social status, and shared symbols of meaning that provided the connective tissue. In this stage, the power of consciousness is used to reflect on feelings of affiliation with others who had certain commonalities—such as living in the same geographic area, sharing ethnic origins, and having a common religion. In a largely preliterate and prerational society, feeling-based communications were the dominant currency of culture. Despite its growing depth, reflective consciousness in the agrarian culture tended to be limited by rigid customs, irrational superstitions, social immobility, widespread illiteracy, a patriarchal society, and the authoritarian character of the church and state. Although all of the basic arts of civilization (such as writing, organized government, architecture, mathematics, and the division of labor) arose during this stage, most people lived as impoverished peasants with no expectation of material change or progress. For the majority, life was brutal, bleak, and short.

While this era represented a dramatic change from the hunter-gatherer and small-village way of life, it also contained many primitive elements—a lack of social mobility, arranged marriages, the oppression of women, restricted access to formal education, and rule by political and spiritual elites.[12] In addition, the range of occupations was quite narrow, as people were expected to pursue the same craft or trade as their family. Although the agrarian era represented a great advance in reflective consciousness and the building of large-scale cultures, it was still only an early stage in the journey to develop the full expression of humanity's capacity for double wisdom.

Stage 4: The era of scientific-industrial civilizations—thinking consciousness. The next great change in consciousness arrived in full force by the 1700s as a number of powerful revolutions blossomed in England, Europe, and the United States. These include a scientific revolution that challenged the belief in the supernatural and the authority of the church; a religious reformation that questioned the role and function of religious institutions; the Renaissance, which brought a new perspective to the arts; an industrial revolution, which brought unprecedented material progress; an urban revolution, which brought masses of people together in new ways, breaking apart the feudal pattern of living; and a democratic revolution, which fostered a new level of individual empowerment and involvement. These powerful revolutions, which still affect us today, were expressions of a new perceptual paradigm and mark a dramatic break with the agrarian era.

When the industrial revolution began in earnest in the late 1700s, more than 90 percent of the population in Europe and the United States lived and worked on farms. Two hundred years later, more than 90 percent of the population in these countries lived and worked in cities and suburbs. In this single statistic is the story of an extraordinary transformation of these societies—a transformation that is now being repeated in countries around the world.

The flat wheel of time that oriented perception in the agricultural era opens up to become a dynamic, three-dimensional spiral in the industrial era. In experiencing that time is "going somewhere," people perceive the potential for material progression or progress. As the mystery of nature gives way to science and an analyzing intellect, material achievements became a primary measure of success.

All of life on Earth has paid a very high price for the learning realized during this era. Although people in industrialized societies are more intellectually sophisticated and psychologically differentiated, they are also more isolated—often feeling separated from nature, others, and themselves. Feelings of companionship and community have been stripped away; many people live nearly alone in vast urban regions of alienating scale and complexity. Unprecedented economic and political freedom have been won, but at great cost when life seems to have little meaning or sense of purpose beyond ever more consumption. The perceptual paradigm of the scientific-industrial era has immense drive but virtually no sense of direction beyond the acquisition of power and things. Despite these limitations, reflective consciousness has advanced considerably in this era, fostering greater citizenship in government, entrepreneurship in economics, and self-authority in spiritual matters.

Stage 5: The era of communication—observing consciousness. Because the communications revolution has enabled humanity to begin observing itself consciously as a species, the capacity for double wisdom is now taking another quantum leap forward. No longer operating largely on automatic, entire societies are increasingly conscious of the simple fact of consciousness—and that changes everything. With reflection comes the ability to witness what is happening in the world and the freedom to choose our pathway into the future.

Reflective consciousness provides the practical basis for building a sustainable future. We cannot afford to run on automatic given the scope and urgency of challenges converging around us. Because of the severity of the combined adversity trends, we are being challenged to pay attention to how we are paying attention collectively. With reflective consciousness we can objectively witness environmental pollution, religious intolerance, poverty, overconsumption, racial injustice, sexual discrimination, and other conditions that have divided us in the past. With a more objective perspective combined with the skills of conflict resolution and the tools of mass communication, we can achieve a new level of human understanding and discover an authentic vision of a future that serves the well-being of all.

Reflective consciousness provides the glue that can bond the human family into a mutually appreciative whole while simultaneously honoring our differences. As we cultivate our capacity for knowing that we know, we begin to heal our sense of disconnection from the larger universe. We catch glimpses of the unity of the cosmos and our intimate participation within the living web of existence. No longer is reality broken into relativistic islands or pieces. If only for a few moments at a time, we see and experience existence as a seamless, living totality. These few moments can have a transformational impact. As the Sufi poet Kabir wrote, he saw the universe as a living and growing body "for fifteen seconds, and it made him a servant for life."[13]

Stage 6: The era of bonding—compassionate consciousness. In the next stage in the unfolding of double wisdom, I believe that an observing consciousness will mature into a compassionate consciousness, and love will genuinely infuse our civilizing activities. The same energy of compassion that binds a family is the unifying force that will make global reconciliation and commitment possible. It is love that will enable us to join in a purposeful union of global scope so as to ensure a sustainable home for all life.

We will not enter this stage with our capacity for compassion fully developed; instead, it will be through working together day after day over generations that we shall evolve our capacity for loving engagement with the world. In turn, it will be the strength of this union that will enable us to withstand the enormous stresses that will be unleashed during the next stage of growth. Where the dispassionate consciousness of the communications era was sufficient to enable us to achieve embryonic reconciliation, it is the compassionate consciousness of the bonding era that will enable us to move ahead to build a creative, global civilization.

A compassionate consciousness would foster a new cultural atmosphere where people feel that they are among friends no matter where they are in the world. During the agrarian era, universal literacy seemed almost impossible, yet we are working to achieve it in industrial societies today. In the same way, a compassionate consciousness as a cultural norm may seem unimaginable today, but it could become a reality within a few generations.

In this era, humanity could take on the restoration and renewal of the biosphere as a common project, which could promote a deep sense of community and bonding. A global culture of kindness could foster world projects that include transforming massive cities into decentralized eco-villages, hosting global celebrations and concerts, and developing ongoing world games as an alternative to warfare.

Stage 7: The era of surpassing ourselves—flow consciousness. With flow consciousness, we experience existence as fresh, alive, and arising anew. The observer no longer stands apart from any aspect of reality, but participates fully. We return consciously to the center of our ordinary lives, and bring the power of our wakefulness to our creative expression. We know, moment by moment, through the subtle hum of knowing-resonance at the core of our being, whether we are living in a way that serves the well-being of the whole.

In this stage, humanity will move beyond maintaining ourselves to surpassing ourselves. In this "surpassing era," compassionate consciousness coalesces into self-organizing action and becomes a force for focused, creative expression in the world. A primary challenge of this era will be to liberate the creative potentials of the species without destroying the foundation of global unity and sustainability developed in the previous stages. To meet this challenge, it will be important to develop and integrate all aspects of reflective consciousness into a balanced whole:

- **Sensing consciousness** of the era of awakening hunter-gatherers
- **Feeling consciousness** of the agrarian era
- **Thinking consciousness** of the scientific-industrial era
- **Observing consciousness** of the communications era
- **Compassionate consciousness** of the bonding era
- **Flow consciousness** of the surpassing era

Figure 8 illustrates these stages in the unfolding of reflective consciousness. When we have fulfilled and integrated the potentials of all these stages, humanity will have become a consciously self-organizing planetary family with the perspective, compassion, and creativity to sustain ourselves into the deep future. We will have consciously developed a rich sensory existence, a textured emotional life, a complex intellectual world, the capacity for reflection and reconciliation, a deep love for the Earth and compassion for all its inhabitants, and the subtle freedom of flow consciousness.

FIGURE 8

Stages in the Unfolding of Civilization and Consciousness

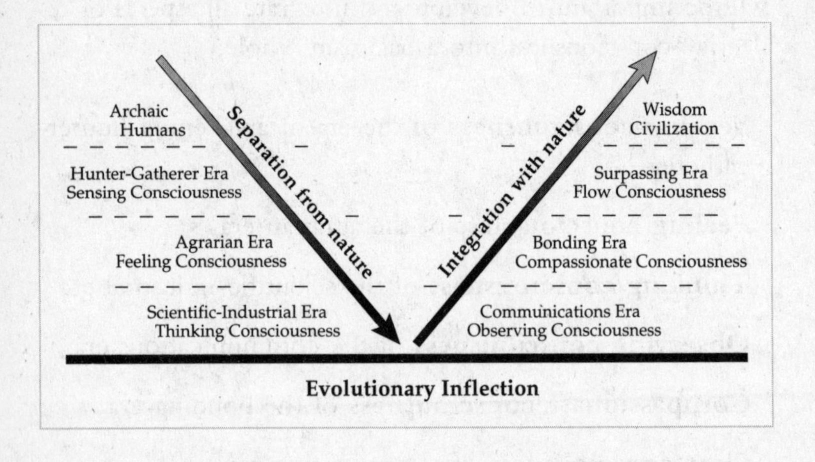

I want to emphasize that *these stages represent a pathway, not a prediction.* What is important is that we begin to see, however dimly, the story of humanity's journey of awakening to the potentials of our self-given name as a species—*Homo sapiens sapiens.*

If my informal surveys of people around the world are correct and humanity is in its adolescent years of development, then we have long-term opportunities in front of us and a long way to go before reaching our full potentials, individually and collectively. The widely shared intuition that we are in our teenage years also suggests that if we are asked, we recognize that we have already traveled a considerable evolutionary distance since our awakening as gatherers and hunters some thirty-five thousand years ago.

While moving toward a globalized world, it is vital that we remember the living wisdom that is at the core of each of the major stages of human experience—as gatherers and hunters, as farmers of land, as dwellers in cities and, increasingly, as dwellers in cyberspace. Each stage represents an invaluable building block in the nested process of developing a sustainable species-civilization that honors both human diversity and global unity.

It is also important to emphasize that this description of the stages in the unfolding of culture and consciousness should not be interpreted to mean that one stage is better than another. As someone who grew up on a farm and now lives in a city, I do not believe that the mind-set of the urban-industrial era is "better" than that of the agrarian era. At the same time, it is important not to romanticize the past. In my view, while all of humanity's past modes of thinking and living will be invaluable for solving the challenges of the future, none of them will be sufficient. We have further learning to do, both as individuals and cultures, in our ways of thinking, living, communicating, and relating.

To portray humans as participants in an experimental and creative universe is not to elevate our stature unduly. Our subordinate status is clear—we do not know how to create a single cubic inch of empty space or matter, let alone the elegance of a flower. The status of the Mother Universe is not threatened by our participation in life on an island called Earth in an immense cosmos with billions of other planets that are likely nurturing life. Rather, we are fulfilling the purpose for which the cosmos was created. We are the agents of evolution, enabling the Mother Universe to reflect upon, and then to surpass herself creatively.

In my view, completing these stages will not represent the end of humanity's journey. It will be our beginning at a new level of possibility. Humans will not become angels or saints; we will simply be awake to the fullness of who and what we already are. Just as reaching adulthood marks the beginning of creative work in the world for individuals, attaining our early adulthood as a planetary civilization will mark the beginning of a new phase in humanity's exploration and learning.

I want to return one last time to the question I asked at the beginning of this chapter: Is there a purpose that is so compelling that it could draw the human family together in common enterprise? My answer is that the highest and most compelling purpose we could possibly imagine is beckoning us—namely, to continue unfolding our capacity for reflective consciousness, both individually and culturally, so that we might learn how to live consciously and compassionately within the Mother Universe. That is the challenge—and the promise—of our journey.

Chapter Nine

Engaged Reflection in the Turning Zone

Mindfulness must be engaged.
Once there is seeing, there must be acting.
Otherwise, what is the use of seeing?
—*Thich Nhat Hanh*

It is not the answer that enlightens,
but the question.
—*Eugene Ionesco*

The Power of Conscious Evolution

WE have looked at the big picture, now what? Responses are clearly called for, but which responses? There are many specific actions that we could take to turn toward a positive future, but where do we begin? These are the kinds of questions that I want to explore in this chapter.

If we human beings are to succeed in our evolutionary journey and mature into our adulthood, I believe that all four of the opportunity trends discussed in chapters three through six need to flourish if we are to succeed. None will be sufficient without the others. All four will be necessary if humanity is to turn the corner and realize an evolutionary bounce. Seeing the universe as alive and a sacred field of evolution will not be sufficient if it does not translate into practical, worldly expressions. Choosing a simpler way of life will not, by itself, be sufficient to secure a sustainable future, although it is surely necessary. Awakening the global brain will not be sufficient if this communication system remains disengaged from the reality of the human situation. Working toward genuine reconciliation across the many barriers that divide humanity will not be sufficient to heal ancient wounds if the process is not supported by change in other spheres of life. In short, achieving an evolutionary bounce will require nothing less than a combined transformation in how we think, live, communicate, and relate.

Despite the great importance of these four factors, there is another that I believe is even more fundamental to our success at this pivotal stage in our evolutionary journey. That factor is the development of the powerful tool of conscious reflection in our personal and collective lives. A basic theme of this book is that evolution is fostering the development of self-organizing systems that have the ability to reflect upon themselves and thereby provide themselves with self-orienting feedback. Building upon this theme, I conclude that the most direct, powerful, and natural way to support ourselves in maturing as individuals and as a species is by increasing opportunities for conscious reflection from the personal to the planetary scale.

Personal reflection refers to seeing ourselves in the mirror of consciousness as individuals and to observing the unfolding of our lives. By analogy, social reflection refers to seeing ourselves in the mirror of collective consciousness as social groupings using tools such as the mass media and the internet. With authentic social reflection, we can achieve a shared understanding and a working consensus regarding appropriate actions. Actions can then come quickly and voluntarily. We can mobilize ourselves purposefully, and each person can contribute their unique talents to the creation of a life-affirming future.

Voluntary or self-organized action will be vital to our success. Our swiftly developing world situation is far too complex for any one individual or group to figure out and propose remedies that will work for everyone. Nonetheless, the world has become so interdependent that our consciousness as citizens needs to match the actual nature of the world of which we are an inseparable part. Thus, this will be a time for rapid learning and experimentation locally while being mindful of how we connect globally.

With sustained reflection and dialogue, we could choose a new pathway into the future. We could look a generation or two into the future and see that we have a choice: we can ignore our situation now and put ourselves through enormous suffering in the future, or we can respond to our situation now and take a higher path. Without reflection, the human agenda seems likely to sink to the lowest common denominator that greed and fear can create.

If we are going to grow up as a species and move the social average of our behavior from adolescence to early adulthood, then the challenge we face is not so much in the world "out there" as it is in our own inner maturity. For example, if in response to adversity trends we focus primarily on making technological and economic changes, we will be missing the heart of the human story—which is making a leap forward to a new level of maturity. No material change is as essential as our collective willingness to look squarely at ourselves and our impact on the Earth—and then to focus our efforts appropriately.

Engaged reflection is the foundation for concerted, voluntary action in democratic societies. In order for social reflection to lead to effective actions, it is important to recognize the creative tension between two pulls toward engagement. On the one hand, there is the pull to have social reflection occur locally where it is tangible, and grounded in face-to-face conversation with others such as family, friends, neighbors, and co-workers. On the other hand, there is a pull to have the scope of social reflection match the scope of the challenges—many of which are at the more impersonal scale of community, country, and planet. Therefore, effective social reflection needs to foster both face-to-face conversations as well as conversations of national and global scale, recognizing there is a creative tension in this polarity.

Small-Group Reflection to Discover a New "Common Sense"

PERHAPS the most important action we can take is to talk with one another, both personally and professionally, about the challenges we face, as well as our visions for a positive future. In living rooms, classrooms, meeting rooms, and boardrooms, we can strike up fresh conversations about humanity's future. Face-to-face conversations can have unexpected power: they can be a vehicle for shared learning that clarifies what we care about, strengthens our commitment to constructive change, and informs what actions are most appropriate for building a sustainable future.[1]

Because the entire world is caught up in a process of globalization, much of humanity is in transition—"between stories" regarding who we are, where we have come from, and where we are going. The human family is just beginning to discover and give voice to our common story that includes, but goes beyond, the stories of our past. It is understandable that we do not have a common story for orienting and organizing ourselves. Humanity has yet to develop a sense of common purpose that mobilizes our collective efforts and elicits our enthusiastic participation in creating a new life together. Because we are all learners together, a core challenge will be to discover our compelling evolutionary story together—a new "common sense," a sense of reality, identity, and social purpose that we can hold in common and that respects our radically changing circumstances. How can we discover such a story in the course of our everyday lives? Let us look at some possible ways.

Study circles. These are typically small groups of a half-dozen to a dozen people who gather together regularly to speak from the heart, tell their stories, and learn from the collective wisdom of the group. Study circles provide a feeling of community around the concerns being explored. Members generally agree to listen respectfully to others without interruption, although they engage in spirited conversations so as to understand the perspectives of those with whom they disagree. Study circles on the theme of voluntary simplicity, for example, have been flourishing in the United States in recent years.[2]

There are innumerable topics to discuss. Here are a few questions drawn from themes in this book: What are people's views regarding the "age" of the human family? What are the circle's views on the major adversity trends? What are views on the major opportunity trends? What do people think is the likelihood that these trends will converge into an evolutionary wall within the coming generation? What might our time of planetary initiation be like? What might an evolutionary crash look like? An evolutionary bounce? Is humanity involved in a larger journey beyond simply maintaining ourselves in ever greater comfort? If so, what is the nature and purpose of that journey? What actions can we take now to live more sustainably and compassionately?

Churches, synagogues, temples, mosques, and meditation halls. Houses of worship could intensify their inquiry into humanity's soulful direction at this critical juncture in history. These spiritual centers could also collaborate with one another in discovering a higher vision that responds to questions such as: What are the guiding spiritual insights for this time of historic transition? What actions could be taken to promote justice and reconciliation in a world divided by gender, wealth, race, religion, geography, and more? What is the role of faith communities as growing numbers of people seek a more experiential or first-hand spirituality? What ways of living are appropriate in this new era?

Classrooms. What is education for if not to learn and think creatively about humanity's future? Educators could develop curricula relevant to the emerging future with its many challenges. For example, reconciliation could become a core area of study; students could learn skills of conflict resolution in the classroom. We could also explore opportunities for lifelong learning to enable people to acquire needed knowledge and skills. From practical skills (such as gardening and health care) to entrepreneurial skills (such as creating a values-based business), schools could make it a priority to educate people to be a self-organizing force for sustainability.

Community and professional groups. Many organizations have local chapters that meet as discussion groups. Conversations about sustainability and the prospects of hitting an evolutionary wall could be integrated into these ongoing groups. Professional groups (such as doctors, lawyers, teachers, and engineers) could explore the nature of a sustainable future in the context of their professions and could encourage their members to take actions appropriate to their professions. For example, engineers may be encouraged to focus on sustainable energy technologies, and lawyers on the legal rights of future generations and other species.

Corporate boardrooms. It is difficult to have a healthy business in a sick world—economic health and environmental health go hand in hand. Instead of focusing exclusively on profits in the short run, corporations could focus on the interests of stakeholders in the longer run. In the past, businesses were able to optimize their performance without worrying about the environmental consequences. Now, with climate change, resource depletion, and other forms of global stress apparent, companies could reframe their approach to business by shifting to triple-bottom-line accounting—and publicly assessing the economic viability, social impact, and environmental consequences of their operations.

There are many places in which to establish the process of social reflection in small group conversations happening daily throughout our lives. In the next section, we turn to consider the other end of this polarity—social reflection at the mass scale of communities, nations, and the Earth.

Social Reflection and Conscious Democracy

WHEN it comes to social transformation, small group conversations and mass communication are two sides of the same coin. Both are essential. Although the foundation for a reflective society is in individual, face-to-face conversations, it is important that these conversations unite at larger scales of social reflection—at the regional, national, and global levels. *Ultimately, the scale of social reflection must match the scale of evolutionary concern—and that is now the entire Earth.*

Communication at these larger scales necessarily involves television and the internet. So, let's look at the circulatory system of effective communication in our modern democracies. In particular, three ingredients are essential to sustain a reflective democracy and learning society: Citizens need to be adequately *informed* through the mass media, they need the opportunity to engage in electronically supported *dialogue* with others in order to build a working consensus, and they need the opportunity to *petition* leaders for making positive changes.

Consider what would happen if any of these three ingredients are missing. If citizens are deprived of essential information, they cannot make sound judgments, so the democratic process will be ill-informed, which is dangerous and counterproductive. If they are adequately informed but cannot peacefully assemble, then they cannot discuss what they know and build a working consensus. The result would be informed individuals with no ability to form the critical mass of consensus necessary for change to occur. Finally, if citizens have the ability to be informed and to gather in electronic dialogue, but are deprived of the ability to convey their shared sentiments to their leaders, then participation is meaningless. The result would be the proverbial "all talk and no action." It is the mutually reinforcing support of these three rights working as a system that provides the foundation for a reflective democracy and learning society.

Currently, around the world, all three ingredients of a conscious democracy are woefully inadequate for meeting the challenge of the combined adversity trends. For the most part, the media treat their viewers as consumers who want to be entertained and distracted, not as citizens who seek to be informed and involved. In addition, we have scarcely begun to exercise the rights of mature citizenship by developing electronic town meetings and other forums where citizens can assemble for sustained dialogue and petition for change. Conscious democracy, with a healthy circulation of communication, is still in its infancy—waiting to be invented by citizen-entrepreneurs facing a time of momentous change.

I recognize that television and the internet are converging into a new medium of global communication, but I find it is useful to focus on television to illustrate three fundamental ingredients of a conscious democracy in the communications era: that it be informed, conversant, and responsive. Let's look at each of these.

An informed citizenry. Many people are ill-informed through no fault of their own. The vast majority of commercial television time is devoted to entertainment, not relevant information. Even news programming is becoming "infotainment." Our situation is like that of a long-distance runner who prepares for a marathon by eating a steady diet of junk food. We are trying to run modern democracies on a diet of televised entertainment just when we are confronting challenges of marathon proportions.

To build a workable tomorrow, we need a quantum increase in the depth and quality of information about our common future. We need a hearty, robust diet of socially relevant media programming about the critical trends and choices facing communities, nations, and humanity as a whole. We need far more documentaries and investigative reports that give us an in-depth understanding of the challenges we face. We need programs that vividly portray what life will be like within a generation if nothing is done to alter current trends, and we need programs that suggest what life could be like if we begin designing ourselves into a sustainable future. Lester Brown, president of the Worldwatch Institute, has written that "while heads of the world's major news organizations may not have sought this responsibility, only they have the tools to disseminate the information needed to fuel change on the scale required and in the time available."[3]

We cannot consciously build a future that we have not first imagined. We are a visual species. When we can see it, we can do it. The power of positive visualization is widely recognized in the realm of individual mind-body medicine, but we have been slow to apply this wisdom to the healing of our social mind-body. Positive visions can be a catalyst for positive actions, which reinforce the power of creative visioning, which can set into motion a self-fulfilling spiral of constructive development.

Here is just one example of the kind of television programming that I believe could stir the public consciousness into authentic reflection. To balance the onslaught of aggressive consumer commercials, alternative commercials, which I call "*Earthvisions*," could be developed. Produced by nonprofit organizations and community groups working in partnership with local television stations, *Earthvisions* could be thirty-second ministories portraying some aspect of a sustainable and meaningful future. They could be low in cost and high in creativity, and done with playfulness, compassion, and humor. They could focus on humankind's connection with the web of life, or on positive visions of the future from the perspective of future generations, or on awakening an appreciation of nature and sustainability. My guess is that the public would be delighted with these refreshing perspectives. Once under way, a virtual avalanche of *Earthvisions* could emerge from communities around the world and be shared over the internet. Other media—such as public access TV, newspapers, radio, and specialty publications—could be used to enrich the dialogue.

We could also develop a rich array of programming beyond thirty-second spots. For instance, television comedies could offer a humorous look at everyday life in a sustainable society (such as the challenges of living in an eco-village). Television news-magazine shows could be developed that focus in-depth on themes pertaining to a sustainable future and our evolutionary success. Dramas could explore the deeper tensions and larger opportunities that families and communities may encounter as we begin designing ourselves into this new future. Because inspiring and hopeful visions for the future are largely absent from our money-obsessed media culture, the opportunities for envisioning creative alternatives are enormous.

A conversing citizenry. Democracy has been called the "art of the possible." However, unless we understand what the majority thinks and feels, we cannot know what's possible. Only when we the public know together what we think as a whole can we confidently communicate a working consensus back to our elected representatives. Therefore, a major step in empowering the citizenry in modern societies is to support communication among citizens so that a working consensus can emerge. A conscious democracy talks to itself and knows its own mind. A conscious democracy is vigilant, watchful, and wide awake. An informed public that knows its own mind can be trusted. After reviewing half a century of polling public opinion in the United States, George Gallup, Jr., found "the collective judgment of the people to be extraordinarily sound."[4] Often, he said, "people are actually ahead of their elected leaders in accepting innovations and radical changes."[5]

Only full and open communication will empower people to act with the level of energy, creativity, and cooperation that our circumstances demand. Initially, people will need to express their anger that the Earth has been so devastated, their disappointments that their material dreams have not been fulfilled, and their unwillingness to make sacrifices unless there is greater fairness. As people work through their anger, sorrow, and fear, they can develop an authentic working consensus. When people can tell their leaders, with confidence, where they want to go and how fast they want to get there, then leaders can do their jobs.

It is time to explore our common future, recognizing that

we are a powerful, contentious, and extraordinary species that has ascended to the verge of a planetary civilization. It is time to begin telling the stories of our long adventure of awakening and to celebrate the sacred nature of our journey.

A feedback citizenry. A conscious democracy requires the active consent of the governed, not simply their passive acquiescence. Of course, citizen involvement is no guarantee that people's choices will always be "right." What involvement does assure is that citizens will be invested in the choices they make. Rather than feeling cynical or powerless, they will feel engaged in and responsible for society and its future.

Once a citizenry knows its own mind and has confidence in its views and values, it can use electronic forums to give feedback to its elected leaders. In a conscious democracy, a working consensus among citizens would presumably guide but not compel elected representatives. Assuming citizen feedback is advisory or nonbinding, it would respect the responsibility of representatives to make decisions and the responsibility of the citizenry to make their views known to those who govern.

My own work to revitalize the conversation of democracy is relevant here. During the 1980s, I launched and directed two nonprofit and nonpartisan organizations—one national called Choosing Our Future and the other local to the San Francisco area called Bay Voice. The purpose of both organizations was to give all people a voice in how the most powerful instrument in our modern society and democracy—television—is being used. Among our many actions, Choosing Our Future worked in partnership with the ABC-TV affiliate in San Francisco to develop a pilot electronic town meeting (ETM) that was seen by over 300,000 people. The hour-long pilot, which aired in 1987, featured feedback from a preselected, random sample of citizens throughout the Bay Area who watched the program from their homes. The program began with an informative mini-documentary to place the issue in context and then moved to an in-studio dialogue with experts and a diverse audience. As questions became better formulated in the studio discussion, the preselected members of the scientific sample were asked to respond. They did so by dialing in numbers on their telephone to register simple votes, which were then tallied by a computer and shown to both in-studio participants and viewers at home. Six votes were taken during the hour-long, prime-time ETM, enabling far more than a one-time "knee-jerk" response from the public. This just begins to suggest the untapped potential for interactivity and dialogue—particularly as these tools are combined with the internet, radio, newspapers, and other forums such as study circles.

Electronic town meetings are an ideal forum for citizen dialogue and feedback. A core issue in building a more reflective society and democracy will be: Who should sponsor these? If ETMs are initiated by commercial television stations and internet sites, they will likely have a bias toward commerce and entertainment. If they are sponsored by a local government body, they will likely be reluctant to deal with concerns of a national and global nature. If they are sponsored by an issue-oriented organization or an institution representing a particular ethnic, racial, or gender group, then there will be a tendency to focus around their concerns. Because of these difficulties, I believe that a new social institution is needed to act on behalf of all citizens as the nonpartisan sponsor of electronic town meetings. Metropolitan areas could develop nonpartisan, nonprofit "community voice" organizations that would perform two key functions: conducting research to determine the issues that are important to the community, and working with television stations, internet providers, and a range of other media to broadcast ETMs. The community voice organization would not promote or advocate any particular outcome; rather, its goal would be to support community learning, dialogue, and consensus building—and then to let the chips fall where they may.

A more specialized form of electronic town meeting also could be developed—a viewer feedback forum. The idea is simple. Television almost never turns its cameras around to look at itself directly. To bring balance, television broadcasters could be held accountable for their programming and advertising in the court of public opinion, perhaps on a weekly basis. A community voice organization could sponsor regular viewer feedback forums. These forums could employ live polling of a random sample of citizens to get an accurate sense of public sentiment. Beyond traditional issues such as TV violence, viewer feedback forums could raise questions vital to a sustainable future for the Earth: Is television creating a level of desire for consumption that cannot be sustained globally? If media-generated desires cannot be sustained, then how will those people who are left out respond? Does the mirror of television accurately reflect the reality of our world? Is the consumerist bias of current programming diverting our cultural attention from critical concerns, dumbing-down our potential, and holding back our evolution? By programming television for commercial success, is the television industry simultaneously programming the mind-set of civilizations for ecological failure? How might the mass media nourish and strengthen our culture and enable us to cope with ecological, social, and spiritual challenges? The working consensus of the community could be presented to representatives of television stations at the end of each program, holding them publicly accountable for their legal responsibility to present a balanced diet of relevant programming that serves the public interest.[6]

The Power of Social Reflection

WHAT difference would it make to have a more conscious society in which there is a healthy circulation of information, conversation, and feedback? Could the process of social reflection be so powerful that it could act as the catalyst for an evolutionary bounce? I believe that it could. Here are five examples to illustrate the kinds of difference that social reflection could make.

Visualizing more sustainable ways of living. I want to repeat what I consider to be a staggering statistic: the average American watches roughly twenty-five thousand television commercials a year. Most are advertisements for a consumerist way of life as much as they are pitches for particular products. What difference would it make to our collective psyche if there were counterbalancing messages in favor of a more sustainable and compassionate world? What if the culture of consumption found itself side-by-side on television with a culture of consciousness that valued simplicity and sustainability? My sense is that high-quality alternative programming such as thirty-second *Earthvisions* and weekly, hour-long viewer feedback forums would awaken public interest in seeing alternative approaches to living and consuming. This highly public feedback process, involving local viewers, would be far more effective in bringing about meaningful change than the cumbersome process of passing laws or involving remote regulatory agencies. Without passing a law, and with breathtaking speed, an entire nation—and ultimately the entire world—could begin reconsidering the levels and patterns of consumption that we value.

Transforming injustice. Martin Luther King, Jr., said that to realize justice in human affairs, "injustice must be exposed, with all of the tension its exposure creates, to the light of human conscience and the air of national opinion before it can be cured."[7] Injustice and inequities flourish in the darkness of inattention and ignorance. But when public awareness is focused on them, it functions as a healing light and creates a new consciousness among all involved. When people know that the rest of the world is watching, a powerful corrective influence is brought into human relations. When economic, ethnic, ideological, and religious violence is brought before the court of world public opinion through the mass media, it encourages people, corporations, and nations to discover more mature and nonviolent ways of relating to others. The global media will soon have the ability to broadcast information about virtually any place, person, issue, or event on the planet within seconds. In a communications-rich world, old forms of political repression, human rights violations, and warfare will be extremely difficult to perpetrate without an avalanche of world opinion descending on the oppressors.[8]

Redefining business success. What is the purpose of business? To serve stockholders by making money? To serve customers by making products they need? To serve the evolution of life on Earth? With the communications revolution, these kinds of concerns are being brought into public consciousness and scrutiny. Business operations are becoming transparent in the new information society. A growing number of advocacy organizations are bringing previously hidden or obscure areas of corporate life into the spotlight of public scrutiny. With no place to hide, firms know that if they are to be viewed and trusted as good citizens of the world community, then they have to behave accordingly. In a communications-rich society, corporations are being pushed to look beyond the interests of shareholders and short-term profits to the well-being of the stakeholders—such as local communities, the environment, and future generations.

Investing in a healthy future. Social reflection could also produce a dramatic increase in investments in a positive future. It is average working-class individuals who own a majority of the massive assets of pension funds and insurance companies. These institutions are currently investing primarily for shorter-term profits in firms (such as tobacco companies) whose products will produce a lower quality of life for retirees in the future (as money will have to be spent to care for more people with lung cancer). We could use study circles at the local level and electronically supported dialogues at the regional, national, and global levels to hold pension funds and insurance companies publicly accountable for investing their funds in companies that will produce a higher quality of life when people retire.[9] Bringing these choices into collective awareness could create a groundswell of public support to encourage some of the largest financial institutions in the world to invest in a healthier future.

Rezoning cities for eco-villages and sustainable neighborhoods. Zoning laws are notoriously difficult to change; yet our changing times call for experiments in different forms of neighborhoods and communities that are designed for sustainability. From community dialogues, a public consensus could emerge to create special zoning districts that would allow new urban forms to grow. If these experimental communities can prosper while serving sustainability and local needs, larger projects could be undertaken, leading to an organic rebuilding of the urban infrastructure. Economic and social life could become more decentralized, creating numerous local anchors of security and sustainability during a time of sweeping global change. Once again, the starting point for this cascade of change is public dialogue and a new social consensus.

These five examples illustrate the rapid and profound impact that authentic social reflection could have in many areas of life—in how we consume, respond to injustice, define business success, invest for the future, and live in our cities. *Social reflection has enormous power for transforming every aspect of life.*

If the power of local dialogues (in study circles, church groups, boardrooms, and classrooms) were combined with the power of electronically supported dialogues across cities, nations, and the Earth, the result would be transformative. No matter what our other differences may be, we all have a common stake in a world that communicates with itself effectively. Whether liberal or conservative, rich or poor, women or men—our future depends on a conscious public that communicates with itself about questions that matter.

As we enter the time of planetary initiation and find our historic moorings gone, social reflection may enable us to reach beneath the surface chaos and discover potentials for collective knowing and communication that were always present but required demanding circumstances to draw them forth. Trusting this sense of shared recognition and knowing, we could simply get it into our collective mind to do things differently. Without an immensity of suffering, we could reflect together on our situation and choose a path of sanity and maturity. As reflection turns to action, our capacity for shared knowing would make possible voluntary actions at the local level that have integrity and coherence at the global level. Trusting in the subtle atmosphere of shared consciousness, people could liberate their creativity locally, mindful of how their actions would contribute to building a sustainable future at the regional, national, and planetary level. A new "spirit of the age"—electric with possibility and invention—could permeate the Earth. People could feel a sense of collective purpose, recognizing they are pioneers of an awakening Earth and stewards for the tender beginnings of a new phase in humanity's future.

There Is Promise Ahead

SOMETIMES when I am looking at the many challenges involved in humanity's evolution, I try to gain perspective by considering what might be happening elsewhere in the universe. Has intelligent life emerged on any of the billions of planets circling around stars like our sun? Has intelligent life successfully evolved through its adolescence to build a sustainable planetary civilization on any of these planets? Astronomer Carl Sagan estimated that there are between fifty thousand and one million civilizations in advance of Earth in our galaxy alone.[10] If there are mature planetary civilizations scattered throughout our cosmos, then the daunting challenge of awakening to early adulthood and achieving a sustainable future must have been met successfully by other civilizations many times before.

Although the awakening of global civilization may be commonplace when viewed from a cosmic scale, the experience of awakening in each world is surely unique. For better or worse, we find our species waking up together on a small planet circling a relatively young star located at the outer edge of a swirling cloud of a hundred billion stars—an average-size galaxy in a universe of billions of galaxies. Will we become one of the unfortunate cosmic seeds that has taken root but is so crippled by fear and self-destructive behavior that we never flower into the fullness of our potential? Or will we become one of the gems of the galaxy, the Earth a place of great beauty and humanity notable for its conscious and compassionate planetary civilization? These are pivotal times for our species and the Earth.

Our coming time of initiation is not a sign of evolutionary failure but of our tremendous success. We are entering a time of great opportunity—and great peril. Although the test of our evolutionary intelligence has already begun, I believe it will be another decade or two before a momentous initiation and great turning will manifest with full force and produce consequences that will reverberate into the deep future. Future generations will look back on the actions we take in these years before the initiation, and will reflect on how we rise to meet the challenge of living through one of the most stressful, pivotal, exciting, and important times in human history.

Because we have the means and the opportunity to

achieve an evolutionary bounce, being alive at this time confers on us all a unique responsibility as well as a unique opportunity. Because we are here now, we are "on duty" and responsible for preserving the evolutionary integrity of the human experiment. *We* are the leaders we have been waiting for. *We* are the social innovators and entrepreneurs we have been seeking. *We* are the ones who are challenged to self-organize and pull ourselves up by our own bootstraps.

We have already traveled far. We are beginning to make the turn back to the Mother Universe and our soulful nature. By my reckoning, we are halfway home and our young adulthood as a species is much closer than we may think. We can get there together. Let's not stop now. There is promise ahead.

For regularly updated information, resources, and links concerning the themes of this book, please see my web page:

www.awakeningearth.org

Acknowledgments

THIS book synthesizes thirty years of research and work. Along the way, a diverse community of individuals has made important contributions to my life and to this book. I would like to acknowledge and thank these treasured souls here: Sherry Anderson, Cecile Andrews, Alan AtKisson, Ted Becker, Barbara Bernstein, Fr. Daniel Berrigan, Juanita Brown, Tom Callanan, Joseph Campbell, Pat Clough, Ram Dass, Dee Dickinson, Dave Ellis, Scott Elrod, Georg Feuerstein, Ellen Furnari, Foster Gamble, Joseph Goldstein, Deborah Gouge, Willis Harman, Barbara Marx Hubbard, Tom Hurley, Jean Houston, David Isaacs, Bob Johansen, Peter and Trudy Johnson-Lenz, Brooks Jordan, Sam Keen, Will Keepin, Marilyn King, Jack Kornfield, Dave and Fran Korten, Roxanne Lanier, Sidney Lanier, Gary Lapid, Ervin Laszlo, Coleen LeDrew, Rob Lehman, Stephen Levine, John Levy, Mike Marien, Robert McDermott, Lester Milbrath, Arnold Mitchell, Brian Muldoon, Michael Murphy, Annie Niehaus, Wendy Parker, Hal Puthoff, Richard Rathbun, Paul Ray, Vicki Robin, Laurance Rockefeller, Rob Shapera, Scott Sherman, David Sibbet, Anne Stadler, John Steiner, Tara Strand-Brown, Brian Swimme, the System Sisters of Perpetual Responsibility, Russell Targ, Mary and Tom Thomas, Sylvia Timbers, Justine and Michael Toms,

Allen Tough, Rinpoche Tarthang Tulku, Sarah van Gelder, Frances Vaughan, Kathy Vian, Roger Walsh, John White, and Tom Yeomans. I have treasured my friendship with Vicki Robin for nearly two decades and I greatly appreciate the enthusiasm and insight that she brought to her introduction of this book. I want to thank Toni Sciarra, my editor at William Morrow, for her talented work as an editor and as a seasoned advocate in guiding this book through the publishing process. I want to acknowledge Deborah Gouge for the skillful and sensitive editing that she has brought to a number of my writing projects in recent years, including this book. Her assistance has been immensely appreciated throughout. I want to thank Coleen LeDrew for being such a supportive partner in the completion of this book as well as for her feedback on the manuscripts. Thanks to Linda Larsen who developed the graphics. Huge appreciation goes to my three sons—Cliff, Ben, and Matt—who brought precious gifts of perspective, balance, humor, and love into my life over the decades in which this book developed. A portion of this book was developed in 1998 as the report "The 2020 Challenge" for a project hosted by the Union Theological Seminary of New York and under the guidance of Deborah Stern and Holland Hendrix.

REFERENCES

CHAPTER ONE: IS HUMANITY GROWING UP?

1. Al Gore, *Earth in the Balance,* NY: Houghton Mifflin Co., 1992, p. 213.
2. Allen Hammond, "Three Global Scenarios: Choosing the World We Want," *The Futurist,* Bethesda, MD, April 1999, p. 43.
3. Joseph Campbell, *The Hero with a Thousand Faces,* New York: Meridian Book Edition, 1956, p. 30.
4. Ibid, p. 37.
5. Peter Farb, *Humankind,* Boston: Houghton Mifflin Co., 1978, p. 431.
6. Ibid, p. 432.
7. Barbara Hubbard, *Conscious Evolution,* Novato, CA: New World Library, 1998.
8. See chapter two of my book *Awakening Earth,* New York: William Morrow, 1993.
9. T. S. Eliot, *Four Quartets,* New York: Harcourt & Brace, 1943.
10. William D. Ruckelshaus, "Toward a Sustainable World," *Scientific American,* September 1989, p. 167.
11. Elizabeth Dowdeswell, "Lessons Learned in Sustainable Development," from a speech, 1998.
12. Kevin Kelly, "Deep Evolution: The Emergence of Postdarwinism," *Whole Earth Review,* Sausalito, CA, Fall 1992, p. 15.
13. Lewis Thomas, *The Fragile Species,* New York: Macmillan, 1992.

CHAPTER TWO: ADVERSITY TRENDS: HITTING AN EVOLUTIONARY WALL

1. Stephen Moore, "The Coming Age of Abundance," *The True State of the Planet,* NY: The Free Press, 1995, p. 110.
2. Fred Smith, "Reappraising Humanity's Challenges, Humanity's Opportunities," in *The True State of the Planet,* p. 379.
3. Julian Simon, *The Ultimate Resource 2,* NJ: Princeton University Press, 1996, p. 12.
4. Ibid.
5. The "Warning to Humanity" was sponsored by the Union of Concerned Scientists, 26 Church St., Cambridge, MA 02238.
6. *Climate Change 1995: The IPCC Second Assessment Report,* Cambridge University Press, 1995.
7. Molly O'Meara, "The Risks of Disrupting Climate," *World Watch,* Nov–Dec 1997, p. 12.
8. Simon Retallack, "Kyoto: Our Last Chance," *The Ecologist,* Nov–Dec, 1997.
9. Personal communication, Donella Meadows, July 1998.
10. William Calvin, "The Great Climate Flip-Flop," *Atlantic Monthly,* January 1998, p. 47.
11. *World Resources: A Guide to the Global Environment: 1996–97,* A publication by The World Resources Institute, The United Nations Environment Programme, The United Nations Development Programme, and the World Bank, NY: Oxford University Press, 1996, p. xi.
12. Ibid., p. 173.
13. The World Bank's mid-range projections are that global population will reach 8.4 billion in 2025 and slightly more than 10 billion by 2050 (Paul Raskin, Michael Chadwick, Tim Jackson, and Gerald Leach, *The Sustainability Transition: Beyond Conventional Development,* Stockholm, Sweden: Stockholm Environment Institute, SEI/UNEP, 1996, p. 21). Recently revised (1996) world population estimates by the United Nations give mid-range projections of 8.04 billion people by 2025 and 9.37 billion by 2050 (reference: *World Population Prospects: The 1996 Revision,* United Nations, forthcoming; presented in the *United Nations Report on the World Social Situation 1997,* New York: United Nations, 1997, p. 14). The Population Reference Bureau's World Population Data Sheet for 1998 projects a world population of 8.08 billion in 2025 (reference: *World Population Prospects: The 1996*

Revision, United Nations, forthcoming; presented in the *United Nations Report on the World Social Situation 1997,* NY: United Nations, 1997, p. 14).

14. Gerard Piel, "The Urbanization of Poverty Worldwide," *Challenge,* Jan–Feb, 1997.
15. Lester Brown, Gary Gardner, and Brian Halwell, "16 Impacts of Population Growth," *The Futurist,* February 1999, p. 40.
16. Joby Warrick, "A Warning of Mass Extinction," *Washington Post,* April 21, 1998.
17. Virginia Morell, "The Sixth Extinction," *National Geographic,* February 1999, p. 46.
18. Lester Brown, "The Future of Growth," in *State of the World 1998,* NY: W. W. Norton, p. 11.
19. Bob Holmes, "Life Support," *New Scientist,* England, August 15, 1998.
20. John Tuxill and Chris Bright, "Losing Strands in the Web of Life," in *State of the World 1998,* p. 42.
21. John Vidal, "Is an Era of Water Wars Looming?" in the *San Francisco Sunday Examiner and Chronicle,* August 20, 1995.
22. Sandra Postel, "Water for Food Production: Will There Be Enough in 2025?" *BioScience,* vol. 48, no. 8, August 1998.
23. Lester Brown and Brian Halwell, "China's Water Shortage Could Shake World Food Security," *World Watch,* Jul–Aug 1998.
24. Paul Simon, *Tapped Out,* Welcome Rain Publishers, 1998 (quoted in *Parade Magazine,* August 23, 1998).
25. Colin Campbell and Jean Laherrere, "The End of Cheap Oil," *Scientific American,* March 1998, p. 81.
26. Ibid.
27. Ibid., p. 82.
28. L. F. Ivanhoe, "Get Ready for Another Oil Shock!" *The Futurist,* Jan–Feb 1997.
29. Anthony DePalma, "The Great Green Hope: Are Fuel Cells the Key to Cleaner Energy?" *The New York Times,* October 8, 1997, D1.
30. Personal communication with Dana Meadows, who is an organic farmer as well as a global researcher.
31. Lester Pearson, quoted in *Changing Images of Man,* O. W. Markley and Willis Harman, eds., NY: Pergamon Press, 1982, p. 13.
32. Source: UNDP, *Human Development Report 1992,* NY: Oxford University Press, 1992.
33. Leslie Shepherd, "44 Million Russians in Dire Poverty," *San Francisco Chronicle,* October 20, 1998.

34. Nicholas Kristof, "Human Costs of Asian Meltdown," *San Francisco Chronicle,* June 9, 1998.
35. Patrick Tyler, "In China's Outlands, Poorest Grow Poorer," *The New York Times,* October 26, 1996, p. 1.
36. Quoted in *World Watch,* Jul–Aug 1998, p. 37.
37. Pacific News Service, "78% of Indian homes without electricity," in *San Francisco Examiner,* March 16, 1997, A14.
38. Lester Brown, *Who Will Feed China,* NY: W. W. Norton, 1995, p. 132.
39. Lester Brown, "Who Will Feed China," *The Futurist,* Jan–Feb 1996, p. 14.
40. Lester Brown, "The Future of Growth," in *State of the World 1998,* NY: W. W. Norton: 1998, p. 12.
41. Ibid.
42. Ibid., p. 13.
43. United Nations Progress Report (5 years after the Earth Summit in Rio de Janeiro), June 1997.
44. For this section I am indebted to historian Clive Ponting who describes how the rise and decline of great civilizations have often been powerfully influenced by environmental factors. Clive Ponting, *A Green History of the World,* NY: Penguin Books, 1993.
45. Ibid, p. 72.
46. Samuel Kramer, *The Sumerians: Their History, Culture, and Character,* Chicago: University of Chicago Press, 1963, p. 268.
47. Ponting, *A Green History,* p. 72.
48. Ibid., p. 76.
49. Ibid., p. 83.

CHAPTER THREE: A NEW PERCEPTUAL PARADIGM: WE LIVE IN A LIVING UNIVERSE

1. Willis Harman, *An Incomplete Guide to the Future,* Stanford, CA: Stanford Alumni Association, 1976.
2. Quoted in David Fideler, "What Is a Cosmos?" from a lecture presented at the Great Lakes Planetarium Association, Grand Rapids, Michigan, October 1995.
3. Lee Smolin, *The Life of the Cosmos,* NY: Oxford University Press, 1997, pp. 252–253.
4. David Bohm, *Wholeness and the Implicate Order,* London: Routledge & Kegan Paul, 1980, p. 175.

5. Michael Talbot, *The Holographic Universe,* NY: HarperCollins, 1991.
6. Louise B. Young, *The Unfinished Universe,* NY: Simon & Schuster, 1986, p. 205.
7. Bohm, *Wholeness,* p. 191.
8. Ibid.
9. Norbert Wiener, *The Human Use of Human Beings,* NY: Avon Books, 1954, p. 130.
10. Max Born, *The Restless Universe,* NY: Harper & Brothers, 1936, p. 277.
11. Brian Swimme, *The Hidden Heart of the Cosmos,* NY: Orbis Books, 1996, p. 100.
12. Renée Weber, *Dialogues with Scientists and Sages,* NY: Routledge & Kegan Paul, 1986, p. 19.
13. Freeman Dyson, *Infinite in All Directions,* NY: Harper & Row, 1988, p. 297.
14. Philip Cohen, "Can Protein Spring into Life?" in *New Scientist,* April 26, 1997, p. 18.
15. Mitchell Resnick, "Changing the Centralized Mind," *Technology Review,* July 1994.
16. Dean Radin, *The Conscious Universe*, San Francisco: Harper Edge, 1997, p. 109. See also Harold Puthoff and Russell Targ, "A Perceptual Channel for Information Transfer Over Kilometer Distances," published in the proceedings of the *I.E.E.E.,* vol. 64, no. 3, March 1976.
17. Radin, *Conscious Universe*, p. 144.
18. Puthoff and Targ, "Perceptual Channel," pp. 338–340. See also R. Targ and H. Puthoff, *Mind-Reach: Scientists Look at Psychic Ability,* NY: Delacorte Press/Eleanor Friede, 1977, pp. 79–83.
19. In the first series of 2,800 trials, the probability of obtaining these results by chance was less than one in three million. In the second series of 1,700 trials (with a more complex configuration of technology), the probability was near chance. In the third series of 2,500 trials (with only the computer but being observed constantly), the probability of obtaining these results by chance was less than one in two thousand. Russell Targ, Phyllis Cole, and Harold Puthoff, *Development of Techniques to Enhance Man/Machine Communication,* report prepared for NASA project 2613, Stanford Research Institute, Menlo Park, California, June 1974. Also see Targ and Puthoff, *Mind-Reach,* pp. 124–129.
20. Freeman Dyson, *Infinite in All Directions,* NY: Harper & Row, 1988, p. 297.

21. John Gribbin, *In the Beginning: The Birth of the Living Universe,* NY: Little, Brown and Co., 1993, pp. 244–245.
22. Ibid., p. 245.
23. Ibid., p. 252.
24. Gregg Easterbrook, "What Came Before Creation?" *U.S. News & World Report,* July 20, 1998, p. 48.
25. Wheeler, quoted in Renée Weber, "The Good, The True, The Beautiful," in *Main Currents,* New Rochelle, NY: Center for Integrated Education, 1975, p. 139.
26. Stephen Mitchell, trans., *Tao Te Ching: A New English Version,* NY: Harper & Row, 1988, chapter 25.
27. The quotation by Shao is taken from Garma Chang, *The Buddhist Teaching of Totality: The Philosophy of Hwa Yen Buddhism,* University Park: The Pennsylvania State University Press, 1971, p. 111.
28. Lao-Tzu, *Tao Te Ching,* trans. Gia-Fu Feng and Jane English, NY: Vintage Books, 1972.
29. A saying of Sojo, quoted in D. T. Suzuki, *Zen and Japanese Culture,* NJ: Princeton University Press, 1970, p. 353.
30. Gospel of Thomas, *Nag Hammadi Library,* James M. Robinson, general editor, San Francisco: Harper & Row, 1977, pp. 129–130.
31. Quoted in Timothy Ferris, *Galaxies,* NY: Stewart, Tabori & Chang, 1982, p. 87.
32. Suzuki, *Zen and Japanese Culture,* p. 364.
33. Francis H. Cook, *Hua-yen Buddhism: the Jewel Net of Indra,* University Park, PA: The Pennsylvania State University Press, 1977, p. 122.
34. Smith, *The Religions of Man,* p. 73.
35. Joseph Campbell, *The Power of Myth,* with Bill Moyers, NY: Doubleday, 1988, p. 217.
36. Wayne Muller, *Sabbath,* NY: Bantam Books, 1998, p. 36.
37. Matthew Fox, *Meditations with Meister Eckhart,* Santa Fe, NM: Bear & Co., 1983, p. 24.
38. Lex Hixon, "The Morning Star of Enlightenment," in Georg and Trisha Feuerstein, eds., *Voices on the Threshold of Tomorrow,* Wheaton, IL: Quest Books, 1993, p. 388.
39. Luther Standing Bear, quoted in Joseph Epes Brown, "Modes of Contemplation Through Actions: North American Indians," *Main Currents in Modern Thought,* NY: Center for Integrative Studies, Nov–Dec 1973, p. 194.
40. Malcolm Margolin, *The Ohlone Way: Indian Life in the San Francisco–Monterey Bay Area,* Berkeley, CA: Heyday Books, 1978.

41. Ibid., pp. 142–143.
42. Ibid., p. 112.
43. The designation of modern humans as *Homo sapiens sapiens* is widespread; see, for example: Joseph Campbell, *Historical Atlas of World Mythology, vol I: The Way of the Animal Powers, Part 1: Mythologies of the Primitive Hunters and Gatherers,* NY: Harper & Row, Perennial Library, 1988, p. 22; Richard Leakey, *The Making of Mankind,* NY: E. P. Dutton, 1981, p. 18; Mary Maxwell, *Human Evolution: A Philosophical Anthropology,* NY: Columbia University Press, 1984, p. 294; John Pfeiffer, *The Creative Explosion: An Inquiry into the Origins of Art and Religion,* Ithaca, NY: Cornell University Press, 1982, p. 13; Clive Ponting, *A Green History of the World,* NY: Penguin Books, 1993, p. 28.
44. Maxwell, *Human Evolution,* p. 111.
45. Mikhail Nimay, *Book of Mirdad,* Baltimore: Penguin Books, 1971, p. 57.

CHAPTER FOUR: CHOOSING A NEW LIFEWAY: VOLUNTARY SIMPLICITY

1. Amitai Etzioni, "Voluntary Simplicity: Characterization, select psychological implications, and societal consequences," *Journal of Economic Psychology,* Elsevier Science, 19 (1998), p. 629.
2. Ibid.
3. Alfie Kohn, "In Pursuit of Affluence, at a High Price," *The New York Times,* February 2, 1999.
4. Richard Gregg, "Voluntary Simplicity," reprinted in *Co-Evolution Quarterly,* Sausalito, CA, Summer 1977 (originally published in the Indian journal, *Visva-Bharati Quarterly,* August 1936).
5. Ibid., p. 20.
6. David Shi, *The Simple Life,* NY: Oxford University Press, 1985, p. 149.
7. Henry David Thoreau, *Walden,* Boston, 1854, p. 168.
8. Thomas Moore, *Care of the Soul,* NY: HarperCollins, 1998, p. 285.
9. George Barna, *The Index of Leading Spiritual Indicators,* Dallas, TX: Word Publishing, 1996, p. 129.
10. Richard Celente, *Trends Journal,* Winter 1997.
11. Paul Ray, "The Rise of Integral Culture," *Noetic Sciences Review,* Sausalito, CA: Institute of Noetic Sciences, Spring 1996.

12. *Yearning for Balance: Views of Americans on Consumption, Materialism, and the Environment,* A report by the Harwood Group about a survey conducted for the Merck Family Fund, 6930 Carroll Ave., Takoma Park, MD (July 1995).
13. Ray, "The Rise of Integral Culture."
14. Ronald Inglehart, *Modernization and Postmodernization: Cultural, Economic, and Political Change in 43 Societies,* NJ: Princeton University Press, 1997.
15. Ibid.
16. Ibid., p. 328.
17. Riley E. Dunlap, "International Attitudes Towards Environment and Development," in Green Globe Yearbook 1994, an independent publication from the Fritjof Nansen Institute, Norway, NY: Oxford University Press, 1994, p. 125.
18. Environics International, news release, Washington, D.C., June 5, 1998, "Citizens Worldwide Want Teeth Added to Environmental Laws."
19. Don Clifton, Chairman of the Gallup Organization, personal correspondence, November 1996.
20. Stuart Hart, "Strategies for a Sustainable World," *Harvard Business Review,* Jan–Feb 1997, p. 71. Another perspective is provided by David Korten, *The Post-Corporate World: Life After Capitalism,* San Francisco: Berrett-Koehler, 1999.
21. Arnold Toynbee, *A Study of History* (Abridgement of vols. I–VI, by D. C. Somerville), NY: Oxford University Press, 1947, p. 198.
22. Ibid, p. 208.

CHAPTER FIVE: COMMUNICATING OUR WAY INTO A PROMISING FUTURE

1. "TV, No Hot Water in Typical Chinese Home," Gallup Poll reported in the *San Francisco Chronicle,* October 27, 1997.
2. Mark Pesce, *Proximal and Distal Unity* (San Francisco, May 1996). From a paper taken from the internet: http://www.hyperreal.com/~mpesce/pdu.html.
3. Joseph N. Pelton, "The Globalization of Universal Telecommunications Services," in *Universal Telephone Service: Ready for the 21st Century?* (Institute for Information Studies, A Joint Program of Northern Telecom and the Aspen Institute, Queenstown, MD, 1991), p. 145.
4. Ibid, p. 156.

5. John Midwinter, "Convergence of Telecommunications, Cable and Computers in the 21st Century: A Personal View of the Technology," in *Crossroads on the Information Highway* (Annual review of the Institute for Information Studies, Aspen Institute, and Northern Telecom, 1995), p. 62.

6. *U.S. News & World Report,* January 6, 1997, p. 60; International Data Group, *MacWorld Magazine,* January 1997, p. 169; Paul Taylor, "Internet Users Likely to Reach 500m by 2000," *Financial Times,* May 13, 1996.

7. Leslie Helm, "A Computer Engineer Shares His Thoughts on the Web of the Future," *Los Angeles Times,* 1998.

8. Robert M. Entman, "The Future of Universal Service in Telecommunications," in *Universal Telephone Service,* p. ix.

9. Peter Russell, *The Global Brain Awakens,* Palo Alto, CA: Global Brain, Inc., 1995, p. 140.

10. Pelton, "Globalization," p. 171.

11. "Fiber to Subscriber," Bell Northern Research, quoted in Pelton, "Globalization," p. 154.

12. Joseph Pelton, "Telecommunications for the 21st Century," in *Scientific American,* April 1998.

13. Lester Brown, et al., *Vital Signs 1998,* NY: W. W. Norton, 1998, p. 22.

14. Payal Sampat, "Internet Use Grows Exponentially," in *Vital Signs,* Lester Brown, et al., Worldwatch Institute, Washington, 1998, p. 98.

15. Alexander Astin, et al., *The American Freshman: Thirty Year Trends, 1966–1996* (Higher Education Research Institute, Graduate School of Education and Information Studies, University of California, Los Angeles, February 1997).

16. Ibid., p. 14.

17. Gene Youngblood, "The Mass Media and the Future of Desire," *Co-Evolution Quarterly,* Sausalito, CA, Winter 1977–78, pp. 12–15.

18. G. Diligenskii, "The Masses and Politics in Modern Russia," *Russia,* no. 1, 1995, p. 73.

19. Quotation from *Future Survey* review on Averting an Electronic Waterloo. CSIS Global Organized Crime Project (Task Force Director and Editor: Frank Cilluffo). Washington, DC: Center for Strategic and International Studies, December 1998.

20. Christopher Bache, *Dark Night, Early Dawn: Steps to a Deep Ecology of Mind,* Albany: State University of New York Press, 2000.

CHAPTER SIX: RECONCILIATION AND THE TRANSFORMATION OF HUMAN RELATIONS

1. Pierre Teilhard de Chardin, *The Future of Man,* NY: Harper & Row, 1964, p. 57.
2. Pitirim Sorokin, *The Ways and Power of Love,* Chicago: Henry Regnery Co., 1967.
3. Jack Kornfield, "The Path of Compassion: Spiritual Practice and Social Action," in *The Path of Compassion,* Fred Eppsteiner, ed., Berkeley, CA: Parallax Press, 1988, p. 29.
4. Sorokin, *Ways and Power of Love,* p. 71.
5. This description is drawn primarily from: Sorokin, *Ways and Power of Love,* p. 67, and Eknath Easwaran, *The Compassionate Universe,* Petaluma, CA: Nilgiri Press, 1989.
6. Gitanjali Kolanad, *Culture Shock! India*, Portland, OR: Graphic Arts Center Publishing, 1994, p. 23.
7. Sorokin, *Ways and Power of Love,* p. 68.
8. Ibid., p. 110.
9. Ibid., p. 69.
10. See, for example, Donella Meadows, et al., *Beyond the Limits*, Post Mills, VT: Chelsea Green Publishing Co., 1992.
11. *Gender and Society: Status and Stereotypes.* An International Gallup Poll Report, The Gallup Organization, Princeton, NJ, March 1996.
12. Susan Davis, quoted in "Women Leaders Review Earth Charter," *Boston Research Center for the 21st Century Newsletter,* Fall 1997, p. 6. See also her article "Principle-Centered Evolution: A Feminist Environmentalist Perspective," in *Women's Views on the Earth Charter,* Boston Research Center, November 1997.
13. Arnold Toynbee, *A Study of History* (Abridgement of vols. I–VI, by D. C. Somerville), NY: Oxford University Press, 1947, p. 555.
14. Donald Shriver, Jr., *Forgiveness in Politics,* NY: Oxford University Press, 1995, p. 7.
15. Desmond Tutu quoted in Terry Tempest Williams, "Two Words," *Orion,* Great Barrington, MA, Winter 1999, p. 52.
16. Archbishop Desmond Tutu, "A Message from the Chairperson," *Truth Talk: The Official Newsletter of the Truth and Reconciliation Commission,* South Africa, July 1998.
17. Dr. Alex Boraine, "A message from the Deputy Chairperson of the TRC," *Truth Talk: The Official Newsletter of the Truth and Reconciliation Commission,* South Africa, July 1998.

18. Michael Battle, *Reconciliation: The Ubuntu Theology of Desmond Tutu,* Ohio: The Pilgrim Press, 1997, p. 39.
19. These examples were drawn in part from Emily Mitchell, "The Decade of Atonement," *Index on Censorship,* May–Jun 1998, London (reprinted in the *Utne Reader,* Mar–Apr 1999, pp. 58–59).
20. John Bond, "Aussie Apology," *Yes! A Journal of Positive Futures,* Bainbridge Island, WA, Fall 1998, p. 22.
21. Ibid., p. 224.
22. Eric Yamamoto, *Interracial Justice: Conflict and Reconciliation in Post-Civil Rights America,* NY: New York University Press, 1999.
23. *Human Development Report 1998,* United Nations Development Programme, NY: Oxford University Press, 1998, p. 37.
24. Ibid.
25. Slobodan Lekic, "Rich Nations Grow More Stingy With Poor Nations," *San Francisco Chronicle,* October 17, 1997, World Section, A14.
26. Glenys Kinnock, "One World," in E. and D. Shapiro, eds., *Voices from the Heart,* NY: Tarcher/Putnam, 1998, p. 122.
27. Desmond Tutu, "Becoming More Fully Human," in Shapiro, *Voices from the Heart,* p. 277.

CHAPTER SEVEN: EVOLUTIONARY CRASH OR EVOLUTIONARY BOUNCE: ADVERSITY MEETS OPPORTUNITY

1. For the designation of modern humans as *Homo sapiens sapiens,* see footnote 43 to chapter 3.
2. Pierre Teilhard de Chardin, *The Phenomenon of Man,* NY: Harper & Row, 1959, p. 181.
3. Ibid., p. 165.
4. Arnold Toynbee, *A Study of History,* NY: Weathervane Books, 1972, p. 132.
5. Clive Ponting, "The Lessons of Easter Island," in *A Green History of the World,* NY: Penguin Books, 1993, p. 168.
6. Ibid., p. 169.
7. Ibid.
8. Jared Diamond, "Easter's End," *Discover Magazine,* August 1995, p. 68.
9. Ibid.
10. Ponting, "Lessons of Easter Island," pp. 6–7.
11. Jared Diamond, "Easter Island Tells Tale of Warning," *San Diego Union-Tribune,* October 25, 1995.

12. Charles Mackay, *Extraordinary Popular Delusions and the Madness of Crowds* (1841; reprint NY: Harmony Books, 1980).
13. Sigmund Freud, *Civilization and Its Discontents,* James Strachey, trans., NY: W. W. Norton & Co., 1961, p. 102.
14. Innocent VIII: BULL *Summis Desiderantes,* December 5, 1484.
15. Kenneth Cameron, *Humanity and Society: A World History*, Bloomington: Indiana University Press, 1973, p. 264.
16. Dean Radin, *The Conscious Universe,* San Francisco: Harper Edge, 1997, p. 293.
17. Alan Weisman, "Gaviotas: Oasis of the Imagination," in *Yes! A Journal of Positive Futures,* Bainbridge Island, WA, Summer 1998, p. 36.
18. Ibid.
19. Donella Meadows, "Village Thrives on Sun, Ingenuity and Spirit," The Global Citizen Column, *Valley News,* March 14, 1998.
20. Adapted from an article by Alan Weisman in the *Los Angeles Times Sunday Magazine,* September 25, 1994. See *In Context,* #42, Context Institute, Fall 1995.
21. Alan Weisman, *Gaviotas: A Village to Reinvent the World,* VT: Chelsea Green Publishing, 1998, p. 8.
22. Ibid., p. 218.
23. Ibid., p. 219.
24. Ibid., p. 222.
25. Pierre Teilhard de Chardin, *The Future of Man,* NY: Harper & Row, 1964, p. 33.
26. Roger Walsh, *Staying Alive,* Boulder, CO: Shambhala, 1984.
27. Václav Havel, President of Czechoslovakia, in an address to a joint session of the U.S. Congress, Washington, DC, February 21, 1990.
28. Ken Wilber, taken from his Website, January 8, 1997.
29. Marianne Williamson, *The Healing of America,* NY: Simon & Schuster, 1997, p. 41.
30. Robert Kenny, "Some Reflections on Group Consciousness and Synergy," NY: International Center for Integrative Studies Forum, April, 1992, p. 4.
31. Ibid.
32. Brian Muldoon, *The Heart of Conflict,* NY: G. P. Putnam's Sons, 1996, p. 167.
33. Ibid.

CHAPTER EIGHT: HUMANITY'S CENTRAL PROJECT: BECOMING DOUBLY WISE HUMANS

1. Thomas Berry and Brian Swimme, *The Universe Story,* San Francisco: Harper & Row, 1992, p. 264.
2. David Bohm has described matter as "condensed or frozen light" and has said that light is the fundamental building block of our cosmos. See also Renée Weber, *Dialogues with Scientists and Sages,* NY: Routledge & Kegan Paul, 1986, p. 45.
3. James Robinson, ed., *Nag Hammadi Library,* 1st edition, San Francisco: Harper & Row, 1977, p. 123. Elsewhere in the Gnostic sources Jesus is quoted by the disciple James as saying: "Search ever and cease not till ye find the mysteries of the Light, which will lead you into the Light-kingdom."
4. Ibid., p. 121.
5. Jesus quoted in The Gospel of Thomas, *Nag Hammadi Library,* p. 124.
6. See, for example, Tsele Natsok Rangdrol, *The Mirror of Mindfulness: The Cycle of the Four Bardos,* E. Kunsang, trans., Boston: Shambhala Press, 1989.
7. Robert Bly, trans., *The Kabir Book,* Boston: Beacon Press, 1977, p. 24.
8. Govinda, *Creative Meditation and Multi-Dimensional Consciousness,* Wheaton, IL: Theosophical Publishing House, 1976, p. 200.
9. John Pfeiffer, *The Creative Explosion,* Ithaca, NY: Cornell University Press, 1982, p. 11.
10. Erich Neumann, *The Origins and History of Consciousness,* Princeton, NJ: Bollingen, 1970, p. 275.
11. For background on the topic of time, see, for example, Marie-Louise von Franz, *Time: Rhythm and Repose,* NY: Thames and Hudson, 1978; Joseph Campbell, ed., *Man and Time,* NJ: Princeton University Press: Bollingen Series, 1957; and J. T. Fraser, ed., *The Voices of Time,* NY: George Braziller, 1966.
12. Ibid., pp. 225–226.
13. Bly, *The Kabir Book,* p. 11.

CHAPTER NINE: ENGAGED REFLECTION IN THE TURNING ZONE

1. Juanita Brown and David Isaacs, "Conversation as a Core Business Practice," *The Systems Thinker,* vol. 7, no. 10, December 1996, Pegasus Communications, Inc., Cambridge, MA.

2. In particular, see the work of Cecile Andrews and her book: *Circles of Simplicity: Return to the Good Life,* NY: HarperCollins, 1997.
3. Lester Brown and Jennifer Mitchell, "Building a New Economy," in *State of the World 1998,* Washington, DC: Worldwatch Institute, 1998, p. 187.
4. George Gallup, Jr., "50 Years of American Opinion," *San Francisco Chronicle,* October 21, 1985.
5. Ibid.
6. How do Earthvisions, Viewer Feedback Forums, Electronic Town Meetings and the rest fit into U.S. communications law (which often provides an important indication of which way the winds of freedom are blowing for the rest of the world)? Does this represent an inappropriate intrusion of the public into the affairs of broadcast TV stations or do these activities represent a fully legitimate exercise of the public's rights and duties in a modern democracy? Because most U.S. citizens are reluctant to act when not given the authority to do so, it is important to know that Earthvisions, Electronic Town Meetings, etc. are fully legitimate expressions of our democratic processes and are strongly supported across a broad spectrum of Constitutional law, Congressional legislation, and FCC regulation.

The legal cornerstone for an electronically supported democracy is found in the First Amendment to the U.S. Constitution, which states that "Congress shall make no law . . . abridging the freedom of speech . . . or the right of people to peaceably assemble, and to petition the Government for a redress of grievances." An electronic town meeting is the fulfillment of this constitutional guarantee. It is a forum where citizens can assemble peacefully and communicate freely with the intention of petitioning appropriate government bodies for changes they think are in the public interest.

Turning from Constitutional to communications law, the public has been given very strong communication rights from the earliest stages in the development of broadcasting law. The predecessor to the Federal Communications Commission—the Federal Radio Commission—in 1928 set down the basic requirement that continues today; namely that broadcasters must give first priority to serving the "public interest, convenience, and necessity." The Commission stated that: ". . . broadcast stations are not given these great privileges by the United States Government for the primary benefit of advertisers. Such benefit as is derived by advertisers must be incidental and entirely secondary to the interest of the public." The Commission further stated that: "The emphasis must be first and foremost on the interest, convenience, and necessity of the listening public, and not on the interest, convenience, or necessity of the individual broadcaster or advertiser."

This high standard of obligation to the public has remained in effect since the inception of broadcasting and is reflected, for example, in the 1969 Supreme Court decision that clarified the responsibilities of broadcasters. The court ruled that: "It is the right of the viewers and listeners, not the right of the broadcasters, which is paramount." In addition: "It is the purpose of the First Amendment to preserve an uninhibited marketplace of ideas in which truth will ultimately prevail, rather than to countenance monopolization of that market, whether it be by the Government itself or a private licensee."

The public has responsibilities just as do the broadcasters. The expressed duty of the public to intervene in broadcasting issues was clearly stated in a major 1966 U.S. Court of Appeals decision: "Under our system, the interests of the public are dominant . . . Hence, individual citizens and the communities they compose owe a duty to themselves and their peers to take an active interest in the scope and quality of television service which stations and networks provide . . . Nor need the public feel that in taking a hand in broadcasting they are unduly interfering in the private business affairs of others. On the contrary, their interest in television programming is direct and their responsibilities important. They are the owners of the channels of television—indeed, of all broadcasting."

It has been thought by some that the sweeping deregulation of television in the 1980s negates this half-century of communications law affirming a duty to serve the public interest. This is not the case. The FCC's 1984 ruling states that "[deregulation] . . . does not constitute a retreat from our concern with the programming performance of television station licensees." Instead, what the FCC has done is to drop specific programming standards and enforcement mechanisms (for example, the FCC dropped minimum requirements for news and public affairs programming, eliminated the requirement that stations keep public logs that show how stations are using air time, and dropped any requirement that the stations poll the community to learn of its needs and wants).

Despite this hands-off approach of the FCC, the broadcasting community continues to recognize it has strong (though now largely unenforced) obligations to serve community interests. For example, in 1985, the President of the National Association of Broadcasters stated: "Broadcasting is indeed a unique industry . . . much different from other corporate citizens in America. . . . We have never advocated removal of the public interest standard. In fact, our obligation is to serve the public interest first and stockholder interest second . . . not the other way around."

7. Quoted in Stephen B. Oates, *Let the Trumpets Sound: The Life of Martin Luther King, Jr.,* NY: New American Library, 1982, p. 226.
8. Joseph N. Pelton, "The Globalization of Universal Telecommunications Services," *Universal Telephone Service: Ready for the 21st Century?* Institute for Information Studies, A Joint Program of Northern Telecom and the Aspen Institute, Queenstown, MD, 1991, p. 146.
9. I appreciate this suggestion given by Nicholas Parker.
10. See the calculations of Carl Sagan and I. S. Shklovskii, *Intelligent Life in the Universe,* San Francisco: Holden-Day, 1966, pp. 409–418.

Index